D0086810

The Human Services Counseling Toolbox
THEORY, DEVELOPMENT, TECHNIQUE, AND RESOURCES

William A. Howatt
Nova Scotia Community College

Brooks/Cole
Thomson Learning™

Australia • Canada • Denmark • Japan • Mexico • New Zealand • Philippines
Puerto Rico • Singapore • South Africa • Spain • United Kingdom • United States

Counseling Editor: Lisa Gebo
Editorial Associate: Susan Wilson
Marketing Manager: Jennie Burger
Project Editor: Trudy Brown
Print Buyer: Stacey Weinberger
Permissions Editor: Susan Walters

Production Services: Johnstone Associates
Text Designer: Penna Design & Production
Cover Design: Cassandra Chu Design
Signing Representative: Ann MacDonald
Compositor: G & S Typesetting, Inc.
Printer: Webcom Limited

Printed in Canada

1   2   3   4   5   6   7   03   02   01   00   99

For permission to use material from this text,
contact us:
Web: www.thomsonrights.com
Fax: 1-800-730-2215
Phone: 1-800-730-2214

Library of Congress Cataloging-in-Publication Data
Howatt, William A.
    The human services counseling toolbox : theory, development, techniques, and resources / William A. Howatt.
        p.   cm.
    Includes bibliographical references and index.
    ISBN 0-534-35932-9
    1. Counseling.   2. Psychotherapy.   I. Title.
    BF637.C6H677   2000
    361'.06—dc21   99-29800

For more information, contact
Wadsworth/Thomson Learning
10 Davis Drive
Belmont, CA 94002-3098
USA
www.wadsworth.com

International Headquarters
Thomson Learning
290 Harbor Drive, 2nd Floor
Stamford, CT 06902-7477
USA

UK/Europe/Middle East
Thomson Learning
Berkshire House
168-173 High Holborn
London, WC1V 7AA
United Kingdom

Asia
Thomson Learning
60 Albert Street #15-01
Albert Complex
Singapore 189969

Canada
Nelson / Thomson Learning
1120 Birchmount Road
Scarborough, Ontario M1K 5G4
Canada

#413200 TO

Contents

This book is dedicated to the late Janet Sears,
a former student, who fought cancer with great dignity
and passed away May 27, 1996.
Her enthusiasm for life has provided me with the inspiration
and energy to continue the journey of writing this text.

Preface

The pressures of being a new teacher probably have not changed much in the past one hundred years—not enough time, not enough resources, and not enough sleep. Having completed my fourth year as a teacher, I can say that there were times when the job of teaching counseling seemed unmanageable. However, my first four years provided me with truly excellent experiences. I attribute this to my students, who always came to school full of enthusiasm for learning.

As a teacher in the new human services program at the **Nova Scotia Community College,** I was unable to find a text that met my particular needs. I was looking for a user-friendly resource to act as a tool in the development of each student's individual eclectic counseling orientation. I was searching for a quick reference guide that included some of the more commonly used counseling techniques, the counseling process, various situations a human services counselor may face, and a paradigm for developing an individual counseling orientation that could help both new and experienced counselors to help their clients.

Although there are hundreds of texts on the market, I was unable to find one for the human services counselor that met the above criteria. I suggested to my class that I would, with their assistance, actually write such a text to meet our particular educational needs, with the goal of publishing a useful counseling resource. When I made the suggestion I already had a good idea of what my students' response would be: "When can we start?"

From the first my students were supportive, never doubting that we would be able to write a quality text. After this experience, I believe more than ever that the self-discipline and cognitive capabilities of students from a community college system are similar to those of university students when it comes to producing a quality product.

I hope this text can be a reminder that community colleges are capable of providing quality. I also hope people who are in positions of authority will re-think old paradigms, and be willing to allow students from a community college the opportunity to explain and demonstrate their knowledge and skill levels. I am confident that, given the opportunity, these students will change many old mind-sets.

I know you will find this text a useful tool in your development as a professional human services counselor. This text is also intended for those who today are already practicing professional counselors.

ACKNOWLEDGMENTS

As an instructor in the human services counseling program at **Nova Scotia Community College** I have had the privilege to work with some very bright, competent, and energetic adult learners. When I began to conceive and construct this text, many of the students were excited by the idea of being a synergetic part of a quality piece of work. If it were not for the energy, assistance, and support of the students, this book might never have been written. It is with the deepest gratitude that I would like to acknowledge each and every one of the following students for their efforts and contributions:

Human Services Counseling Class 1994–95

J. A. Sears

D. C. Borden-Stevenson	R. M. Gaudet	E. C. Lloyd	D. B. Porter
C. Stromenberg	A. Graves	J. N. Staples	B. F. Reid

Human Services Counseling Class 1995–96

Duncan Bremner	D. Stephen Fraser	Jennifer Rogers	Jennifer Bigelow
James Meister	Nadine Whalen	Allyson MacInnis	Frank Forsythe
Nicole Belliveau	Janey Carter	Karen Schlogel	Vicki Morrison
Valerie Peterson	Bernadette Burns	Heather Dolan	Paula Sheppard
Karen Lewin	Joan France		

I would further like to thank the following people for their assistance: John Colville, Mary Ellen Crouse, Tanya Lloyd, Tracey Risser, and Scott Sears for their expertise in graphic design; Al Kingsbury and Gary Dunfield, for their expertise and assistance in the area of desktop and traditional publishing; B. Ross, for his meticulous proofreading of the text, and excellent suggestions; S. Fraser, M. Kougell, and A. McCordick, for proofreading the text for technical accuracy; S. Black, pharmacologist, for his ideas and suggestions for the addictions intervention chapter; Ken Pierce, for his expert opinions on the reality therapy and NLP chapters; R. Bustin, general practitioner, for proofreading the eating disorders chapter for medical accuracy; and N. Porter-Steele, for accuracy in the chapter on transactional analysis.

Many thanks to each of these individuals for their cooperation and enthusiasm when asked to proofread the text. I truly appreciate your commitment to both community and education.

In addition, I am indebted to the following colleagues, who reviewed this text: Diane Coursol, Mankato State University; Rob Lawson, Western Washington University; Beth Nimmo, Metropolitan Community College; Lenore Parker, Creative Solutions; Judith Slater, Kennesaw State College; and Willie Whitfield, Ivy Tech State College.

Introduction

The purpose of this book is to provide you, the human services counselor-in-training, with a toolbox that you can use to design your own counseling orientation. As you will come to see, this is a very personal task. No two orientations are the same, just as no two individuals are the same. The counseling orientation is the result of careful study of an array of theories and therapies followed by a process of integrating your own individual preferences with those of one or (more likely) a number of therapeutic approaches.

To help you develop such an individual, integrative counseling orientation, I have incorporated some visual tools here. One is a **continuum** (an X-Y axis) that places each orientation in terms of how directive it is. The other is a **mind map**, which graphically portrays the particular therapy under discussion.

This text is a guide designed to help you begin your professional development as a human services counselor. Think of this book as a journey that you can take toward your destination of a personal counseling orientation. In this journey, there are three "stops" on the way to producing this orientation.

Part One

Part One explores the foundations upon which professional counseling rest. It presents the basic concepts that every counselor needs to know. This section is not intended to be complete, but to be used as a resource. It can never replace formal training under supervision. However, it will provide you with an array of options. Following is a chapter-by-chapter preview of topics.

Chapter 1 is a comprehensive introduction to human services counseling, highlighting what a new counselor must consider before beginning to work with a client.

Chapter 2 provides a user-friendly checklist of professional ethics that a counselor must embody.

Chapter 3 focuses on multiculturalism, providing an overview of the aspects of counseling that are touched by differences between client and counselor. We must always remember that we are dealing with people, that all individuals are different, and that it is essential to learn to work sensitively so that we create a sense of ease in the therapeutic setting.

Chapter 4 presents a professional assessment for client readiness. No matter what orientation you have developed, a client will not be responsive unless there is motivation for change. Readiness for counseling is thus critical to the success of the counseling process. When the client is ready, we can assist in bringing about the desired changes.

Chapter 5 provides a model of assessment and treatment planning that is both thorough and practical.

Chapter 6 includes an instrument that can alert the counselor to possible addiction concerns. Because the presence of addiction complicates the counseling situation, it is essential that the counselor be aware of it from the outset.

Chapter 7 turns to the process of developing your own counseling orientation and introduces the material of Parts Two and Three.

Part Two

Here you will find a variety of counseling theories and techniques from which you can choose for your own counseling orientation. Twelve commonly used perspectives are presented with their accompanying techniques. Chapters 8 through 19 range from Carl Jung to William Glasser and Aaron Beck, and include topics from existentialism to family systems. Among these distinguished therapists you will find an array of approaches that will suit your personal style to a greater or lesser degree. As you complete the section, it will become clear that certain theories feel compatible and can be integrated into your own orientation.

Part Three

First Aid Counseling offers a variety of topics to prepare you for the challenging situations that you will face from time to time as a human services counselor. Explanations, definitions, and interventions for some of the more common difficulties are presented here. It is important that you include suicide intervention in your orientation. I would also underscore the importance of addressing crisis management and counselor burnout. Depending on your work setting or the primary population you are addressing, other topics of Part Three will be applicable as well.

Appendix

The final section of this text contains examples of my students' development of their own counseling orientations. You will see the mind maps that some of the students created, accompanied by brief write-ups of their integrated orientations. When students complete my course, they must take responsibility for knowing how and what they are going to do when they begin human services counseling. In working with other human beings, we need to ensure that we are as prepared as possible.

Part I

Building Blocks of a Counseling Orientation

Chapter 1

An Introduction to Human Services Counseling

Human services counselors are individuals who work (paid or unpaid) in the human services field, helping people who require their assistance.

Human Services Class, 1995

The purpose of this text is to provide those of you within the human services field an easy-to-use reference for developing individual counseling orientations—a useful tool for your professional "toolbox." In contrast to texts weighted heavily with theory, this one has been put together based on the KISS principle: *Keep it simple. . . .*

My overall philosophy is based on the premise described by Rotter (1982) as *internal locus of control,* that all behavior is attributable to individual choices. Professionals who agree with the concept of internal locus of control may also want to consider the suggestion offered by Dawson (1993), who states that "letting the clients do the work is of paramount importance." As a human services counselor, I agree with Barrett (1991), who teaches that, no matter what any counselor does, in the end clients will always do what they want. This means the counselor cannot make clients do what they do not want to do; counselors are only able to control their own behavior.

Professional counselors who operate from a position of internal locus of control may find many of their clients operating from the opposite end of the continuum, which Rotter (1982) called *external locus of control.* This means they believe that outside factors affect or control human behavioral outcomes. When counseling begins, if the counselor and the client are not working from the same paradigm, difficulties may ensue. Human services counselors need to be cognizant of this potential difference in perceptions and formulate their counseling orientation in a way that addresses this issue.

Workers in the field of human services will be dealing with a wide variety of clients who have diverse concerns (Table 1–1). Whatever their concerns, the same question can be asked by the counselor: How can I address this client's concerns? Before professionals address any client issues, they must first obtain a clear understanding of their own role in their particular human services agency. If part of your

Table 1–1	Client Concerns	
Self-discipline	Self-worth	Low self-esteem
Family dysfunction	Self-concept	Motivation
Career change	Loneliness	Substance abuse
Sexuality	Violence	Education
Eating disorders	Displacement	Cultural issues
Gambling	Relationships	HIV/AIDS
Grief	Abuse	Anger

job responsibility is to counsel clients, it makes sense that you acquire some counseling skills. The sum of your skills, both acquired and innate, will provide a foundation for your own integrative counseling orientation. As counseling experience and knowledge are developed, your personal counseling orientation will come to include a number of counseling techniques. The sole purpose of this text is to guide you as a human services counselor on the journey toward focused, consistent, and effective counseling.

WHAT IS A HUMAN SERVICES COUNSELOR?

In the therapeutic field there are many levels of training, both experiential and academic, as well as numerous methods of certification, from licensing to academic degrees and other higher levels of specialized training. Professionals are cautioned to know the limitations of their training and certification for therapeutic practice. For the purposes of this text, a human services counselor is an individual in the helping professions who works with people in a variety of settings including group homes, government and private agencies, and private practice. Because of working directly with the client, the human services caregiver is inevitably involved in counseling. Thus, I define **human services counselors** as those who provide counseling as their main function, although it may be only one of their roles with the client. Training in counseling may or may not have come from traditional colleges or universities, such as graduate schools of psychology or education.

In preparing for counseling responsibly as a professional in the human services field, you will need to consider what is meant by counseling, what the goals of the human services counselor are, and what your own limitations are. In my opinion, if you do not develop your own counseling orientation you risk increasing the distress of clients because you will lack direction and focus.

Some Definitions of Counseling

1. Counseling is to help clients see that they can control their internal actions and experiences more than they first think (Egan, 1994).

2. Counseling is a facilitative and collaborative process that involves supporting and empowering people on their path to personal growth and well-being (Human Services Counseling Classes, 1995, 1996).

3. Counseling can best be defined as one person, the counselor, helping another person, the counselee, who wants the help (Glasser, 1994).

4. Counseling has been considered so intricate, with each word, gesture, inflection, and silence being considered significant, that it could only happen between a skilled counselor and a willing counselee (Myrick, 1987).

5. Mutual exchange of advice, options. . . . (*Funk and Wagnalls Standard Desk Dictionary,* 1979).

You have doubtless noted some differences in the above definitions. This underscores the importance of formulating a personal definition of counseling. Once you are able to do this, you have positioned yourself to begin creating your own counseling orientation.

Some may ask what the difference is between counseling and psychotherapy (Table 1–2). Reviewing the literature reveals that this arouses a great deal of debate. There is no universally agreed upon definition for either counseling or psychotherapy. Both approaches rely on psychological theory, treatment plans, and assessment. From my perspective, they can be seen as a continuum, with one end of the continuum as counseling and psychotherapy as the other.

Belkin (1988) provides the following research point to consider as you formulate your definition of counseling:

• What distinguishes counseling from psychotherapy . . . is the degree of the client's disturbance. In counseling, the client is an adequately functioning individual, but in psychotherapy, the patient is neurotic and pathological.

• Hansen, Stevic, and Warner (1977) state while they [counseling and psychotherapy] appear at opposite ends of the continuum, they are neither dichotomous nor mutually exclusive ways of helping people in need. Nevertheless, the counselor and psychotherapist generally operate at different ends of that somewhat mystical continuum.

Table 1–2 *Differences Between Counseling and Psychotherapy*

Counseling	*Psychotherapy*
Works at a conscious level	Works at both conscious and unconscious levels
Purpose is to guide change	Purpose is to guide changes, as well as direct change
Acts as a resource for change	Acts as the change agent, as well as a resource
Usually short-term interventions because problems are not ingrained in personality	Can be both short-term and long-term, addressing change of personality and psychopathology

I can assure you that in the world of counseling and psychotherapy many have created clear-cut definitions. I believe that as a counselor grows with experience, expertise, and certification of competencies the gap between the two gets narrower. As a new human services counselor, I recommend you err on the side of caution and be ever aware of your boundaries.

Goals of the Human Services Counselor

The goal of a counselor in the human services counseling field is to assist the client to solve, reduce, limit—or in some cases eliminate—behaviors that are causing functional impairment. When clients want to change some component of their life, the counselor's role is to help them make the change. *No matter what the education or skill level of counselors, they can only act as a guide to change. The counselor cannot make the change for the client.*

Because they are employed by substance abuse agencies, alternative housing organizations, crisis centers, group homes, work centers, corrections, and the like, human services counselors usually work on the front lines with clients. They are positioned to develop a trusting therapeutic relationship that can be vital to the health and function of their clients. Those with strong counseling skills will significantly enhance their clients' chance of achieving desired changes. As a professional, it is important to define your goal of service and to pursue the appropriate education and certification for that field.

Limitations of the Human Services Counselor

From the outset, it is the responsibility of human services counselors to work within the guidelines of their particular agency and their professional organizations, adopting professional standards in determining counseling boundaries. Having noted their professional boundaries, human services counselors must then become aware of personal counseling limitations. When in doubt, new counselors need to consult with their professor or supervisor.

ASPECTS OF A COUNSELING ORIENTATION

Every professional's counseling style is likely to be eclectic, meaning that it may contain elements borrowed from more than one theory, because it will be integrated with the counselor's own personality and beliefs. However, it is imperative that counselors be grounded in at least one theory about which they feel comfortable. Jacobs, Harvell, and Masson (1994) suggest that those who do not have a good working knowledge of at least one theoretical perspective often do not go beyond the surface interactions and sharing. In other words, lack of an adequate theoretical foundation may significantly impair a counselor's ability to perform the job competently.

As professionals, we support clients through our individual counseling orientations as they seek to reinvent themselves. Boffey (1993) states: "When you reinvent yourself, you create new beliefs, values, assumptions, and principles and develop a way of living which unifies your behavior around these new foundations."

The following three components need to be taken into consideration as you develop your counseling orientation:

1. *Social norms* relating to the vocation.
2. The characteristics of the *population* from which your clients come.
3. Your *personal ethics* relating to the vocation.

Social Norms

It is imperative that professionals work within the boundaries set by the society in which they live. Society expects and requires that we do not inflict pain on our clients, whether physical or emotional. For example, I live in Canada, where we have laws and regulations pertaining to various forms of abuse (the Canadian Criminal Code) and ethical violations defined by the Canadian Psychological Association's Code of Ethics. We are called upon by society and professional organizations to respect the client's humanity, treating the client as we ourselves would like to be treated in a similar situation. It is the personal and professional responsibility of professional counselors to be up to date and knowledgeable about the laws and professional standards of their community.

A human services counselor is a *service provider,* a person who helps, assists, and contributes to the well-being of others. This does not involve controlling the lives of our clients, although it may mean planning some of their daily activities. When clients develop progressive difficulties with their behaviors, the human services counselor is more like the carpenter who helps to repair a house than the architect who redesigns it. Our counseling orientation, whatever its strengths, must reflect the paradigm of services to our clients rather than that of control over them. Peck (1994) describes the counseling paradigm in this statement: "My purpose, therefore, is not to make people or organizations more pain-free, but to assist us organizational creatures to be healthier, happier, and more alive."

Target Population

As a professional, it is important to know your target population. Some relevant factors include age, demographics, gender, race, culture, religion, socioeconomic status, and even substance abuse. All of these areas of consideration will provide insight into the client's environment. Then, based on a personal counseling orientation, the human services couselor can choose a counseling direction that is individualized to suit the client.

Richard Bolles, well-known author of the perennial *What Color Is Your Parachute?,* described in his book *The Three Boxes of Life: And How to Get Out of Them* (1981) a philosophy that can be modified and used by the human services counselor. He suggested the following checklist for a counseling orientation:

1. *Analyze the surroundings.* As a professional, you must first find an area of interest and explore what it will demand from you.
2. *Keep afloat.* Once you have decided what population and concerns you want to work with, you must then determine what counseling orientation will fit both your personality and area of choice.
3. *Search for a purpose.* Once you have decided on the field you want to work in, and have started to develop your counseling orientation, you will want to

develop personal goals regarding what you want to be able to achieve. This will help you sort out what your ultimate goal is as a human services counselor.

4. *Strive for improvement.* It may also be important to consider periodically how much success you actually achieve with your present eclectic counseling model. If you are not able to obtain what you believe are acceptable outcomes, you may want to consider more training. Professionals should always be evaluating their productivity and continuing to grow professionally (e.g., through workshops, educational upgrading, consulting with peers in the field).

PROFESSIONAL AND PERSONAL ETHICS

It is essential that human services counselors review both personal and professional ethics while developing their counseling orientation and *before they work with a client.* I request all my students to reflect deeply upon the following formula:

professional ethics (PE) + personal ethics (PE) + counseling orientation (CO) + organization (O)

= what and who we counsel

To explain further, the formula includes the following elements: (a) the standards set forth in the code of ethics of your professional organization; (b) your own moral principles, which underlie your choices regarding both actions and limitations; (c) your counseling orientation—the theoretical models you have chosen that fit both your personality and your client population and allow you to be as congruent as possible; (d) your organization, which refers to where you are working, and the policies, procedures, and standing orders that define your role there. Putting all these elements together will clarify your boundaries as a professional counselor.

The *Webster's New World Dictionary* (1994) defines the word *ethic* as a system of moral standards or values. An excerpt from the code of ethics of the National Board of Certified Counselors (1996) states that ethical behavior among professional associates (both certified and non-certified counselors) must be expected at all times. When accessible information raises doubt as to the ethical behavior of professional colleagues, whether certified or not, the certified counselor must take action to rectify the situation. Guidelines for professional conduct established by any group, body, or association are called its *code of ethics.* Thompson points out (Gilliland and James, 1996) that ethical standards have no legal bearing, although they may closely parallel the law. Overall, no code of ethics can outline the appropriate course of action for every situation. Ultimately, professionals are expected to exercise prudence when interpreting and applying ethical principles to specific situations.

Within the framework of human services, counselors have to choose the degree to which they will be *directive* in their style. Some people are more comfortable with leading, advising, or suggesting than with confronting. Because different theories are connected with different styles of directing, with some advocating that counseling be straightforward while others require a more subtle approach, it is important to know what style fits you best.

First, ask yourself this question: *To what extent do I have the right to give direction to others?* (Remember that this is in the context of the professional as a provider of human services.) Think about the question seriously, because your answer can be of great help in choosing an orientation. It is important to consider the flip side as well: *To what extent do I have the right to withhold active influence of clients?* This second question raises the issue of the degree to which you believe you are responsible to the client. Schultz (1994) states that the ethical ideal of competence reflects the expectation that professionals in the counseling field must possess an ability to evaluate their own skills honestly and objectively, and to make reasonable, responsible decisions of both a professional and moral nature.

The American Counseling Association (1995) states in its Code of Ethics, ". . . counselors are aware of their own values and beliefs and behaviors, and how these apply in a diverse society, and take care to avoid imposing their beliefs on clients." Self-evaluation, self-exploration, and awareness are an essential part of building your personal ethics. It is essential to be able to explore your character and assess your honesty with clients about the influence of your personal values. Your ability to remain objective in matters pertaining to your client will enhance your relationship with clients and enable their growth. Howatt (1996) states that counseling is a process whereby clients are challenged to evaluate their lives honestly and then decide for themselves how they will modify their values and behaviors. Finally, as stated in the newly revised (1995) Ethical Principles of Psychologists and Code of Conduct of the American Psychological Association (APA), where differences of age, gender, race, ethnicity, national origin, religion, sexual orientation, disability, or socioeconomic status significantly affect psychologists, it is necessary to obtain the training, experience, consultation, or supervision to ensure the competency of their services, or to make appropriate referrals.

Counseling Orientation

The Code of Ethics of the National Board of Certified Counselors (1996) states that "the primary obligation of counselors is to respect the integrity and promote the welfare of a client." The American Counseling Association states that "in the counseling relationship, counselors are aware of the intimacy of the counseling relationship, maintain respect for clients, and avoid actions that seek to meet their personal needs at the expense of the clients (Code of Ethics, 1995). When working in groups, counselors must be willing to examine both their ethical standards and their level of competence. The orientation of a counselor is a combination of theory, practice, beliefs, and personality. As a counselor you will gravitate to, and should have a thorough understanding of, at least one theory. Yet it is the norm for counselors to incorporate more than one theory, practice, or technique into an orientation, and this is called the eclectic approach. Be aware of the importance of social norms when considering your orientation. Society expects and requires that we do not inflict pain upon our clients, whether physical or emotional. We must also consider our responsibilities to the client, and what theory fits our belief system. Schulz (1994) states that the ethical ideal of competence reflects the expectations that professionals in the counseling field must possess an ability to evaluate their own skills honestly and objectively, and make reasonable, responsible decisions of both a professional and moral nature.

Organization/Agency

Incorporated into all of the foregoing must be the code of practice, guidelines, or other policies of the agency for which you work. In some cases, these codes, guidelines, or policies may conflict with personal ethics or orientation, creating a personal ethical dilemma. Is it ethical to work for an agency or organization whose policies are in conflict with your own personal ethics? A self-evaluation will be necessary to answer this and many other ethical questions.

Summary

This text is based on the presupposition of an internal locus of control—namely, that the main responsibility for behavior lies within oneself. This provides a key to the definition of counseling. It is important to know how counseling is distinguished from psychotherapy, what your professional goals are, and what are the limits of human services counseling.

In developing your own counseling orientation, you need to consider the social norms relating to the vocation, your client population, professional ethics, and your organization or agency. A checklist for anyone intending to enter the field of counseling includes analyzing the surroundings, developing a suitable counseling orientation, developing personal counseling goals, and continually considering how to improve personally and professionally.

References

Barrett, C. (1991). *Beyond AA: Dealing responsibly with alcohol.* Greenleaf, OR: Positive Attitudes.

Belkin, G. C. (1988). *Introduction to counseling* (3rd ed.). Dubuque: Brown.

Boffey, B. (1993). *Reinventing yourself: A control theory approach to becoming the person you want to be.* Chapel Hill: New View.

Bolles, R. N. (1981). *The three boxes of life: And how to get out of them.* Berkeley: Ten Speed.

Corey, G. (1995). *Theory and practice of group counseling* (4th ed.). Pacific Grove, CA: Brooks/Cole.

Corey, G. (1996). *Theory and practice of counseling and psychotherapy* (5th ed.). Pacific Grove, CA: Brooks/Cole.

Corsini, R. J., Wedding, D. (Eds.) (1989). *Current psychotherapies.* Itasca, IL: F. E. Peacock.

Dawson, B. (1993). *The solution group: Self-evaluation through group counseling.* Chapel Hill: New View.

Egan, G. (1994). *The skilled helper: A problem management approach to helping.* Pacific Grove, CA: Brooks/Cole.

Funk and Wagnalls Standard Desk Dictionary (1979). New York: Author.

Glasser, W. (1994). *Control theory manager.* New York: Harper & Row.

Howatt, W. (1996). *Counselling for paraprofessionals: Formulating your eclectic approach.* Halifax, NS: Nova Scotia Community College Press.

Jacobs, E., Harvill, R., Masson, R. (1994). *Group counseling.* Pacific Grove, CA: Brooks/Cole.

Kral, M. J., Enns, K. L. (1990). Handbook for psychologists and psychological service providers. *Manitoba Psychologist 9:*49–66.

Myrick, R. (1987). *Development guidance and counseling: A practical approach.* Minneapolis: Educational Media Corporation.

National Board of Certified Counselors (1996). *Code of Ethics.* Available: http:www.nbcc.org/ethic.htm

Peck, S. (1994). *A world waiting to be born: Civility rediscovered.* New York: Bantam.

Rodriquez, G. (1994). *Pocket criminal code.* Toronto: Carswell.

Rotter, J. B. (1982). *The development and applications of the social learning theory: Selected papers.* New York: Praeger.

Schultz, W. (1994). *Canadian ethics casebook.* Ottawa: Canadian Guidance and Counseling Association.

Scissons, E. (1993). *Counseling for results: Principles and practices of helping.* Pacific Grove, CA: Brooks/Cole.

Webster's New World Dictionary, 3rd College Edition (1994). Toronto: Prentice-Hall.

Wubbolding, R. (1986). *Using reality therapy.* New York: Harper & Row.

Chapter 2

Professional Ethics Checklist

IMPACT OF COUNSELOR VALUES ON ETHICS

Although we are all products of our environment (and thus influenced by our parents, peers, and society in general), it is possible within the therapeutic setting to remain objective and refrain from forcing our personal beliefs on our clients. In the counseling process clients are challenged to look at their values honestly and then decide for themselves in what ways they will modify these values and subsequent behavior. It is important that we be aware of our personal values and beliefs, how they were instilled, and how they influence our interventions with clients. Self-exploration is a vital part of the counseling process. Here are a few questions for you to reflect upon:

1. Do counselors always need to agree with their client's belief system?

2. Do counselors always need to agree with their agency's values, at the risk of compromising their own?

3. Do counselors have the right to insert their own ethics in the client's situation? If so, what is appropriate?

4. Do counselors need to overlook the client's values to avoid judging the client as they try to help?

I believe counselors must explore their own values and beliefs. In essence, we are talking about personal character: what is it, and what do we stand for? For those who are not sure, I recommend Stephen Covey's (1989) *7 Habits of Highly Effective People* to learn how to develop personal habits that build the essence of your character.

ETHICAL AND LEGAL ISSUES

All professionals must become familiar with community, regional, and national laws as they pertain to the profession. The difficulty for the human services counselor lies in integrating the law with ethical standards.

Professional codes of ethics establish standards; they provide a basis for professional accountability, protect clients from unethical practice, and provide a foundation for improving professional practice. It is important not to confuse legal and ethical behavior, although obeying the law is part of ethical behavior. Ethical standards have no legal bearing, although they may closely parallel the law (Gilliland & James, 1996); instead, they are general guidelines on conduct for a particular profession. Breaches of standards may not be addressed in courts of law; they will, however, be addressed by the individual's professional association. The following are areas generally covered by ethical standards: competency of the counselor (reasonableness, prudence); relationships with clients; confidentiality; and due process (which is related to competency and reasonable action, in that all actions you take must be justifiable as reasonable actions of a competent practitioner). The best protection against malpractice and liability is to respect the client, keep client welfare paramount, and practice within the framework of professional codes. Corey (1996) sums up the adaptation to ethical codes as follows:

> Even though codes are becoming more specific, they do not convey ultimate truth, nor do they provide ready-made answers for the ethical dilemmas that practitioners will encounter. Ultimately, professionals are expected to exercise *prudent judgement* [italics added] when it comes to interpreting and applying ethical principles to specific situations. No code of ethics can outline the appropriate course of action for every difficult situation.

Sexual Relationships

There is no room for ambiguity in the relationship between counselors and clients. Sexual relationships with clients are unethical, and it is the responsibility of the counselor to have no sexual contact with the client. Even after the therapeutic relationship has been concluded, professionals disagree on the amount of time that must elapse before a relationship is ethically acceptable. I am of the belief that it is never ethical to have a sexual relationship with a client or former client.

Nonsexual Dual Relationships

The counselor would be wise to follow the boundaries laid out by the Addiction Intervention Association (AIA), as detailed in their newsletter *The Beacon* (1996):

The counselor should refrain from:
- Buying/selling a product to/from a client
- Entering a business or financial arrangement with a client
- Attending a client's social event
- Inviting a client to a social event
- Developing a friendship or social relationship with a former client
- Accepting a gift from a client
- Counseling a client's close friend, family member, or partner
- Counseling an employee, trainee student, or one of their own friends

- Asking a client to employ, or aid the employment of, their friends, family
- Referring a client to a close friend or family member
- Exploiting any information given by the client outside of the therapeutic relationship, for financial gain
- Revealing a client's identity, without his/her written consent, for social or professional gain
- Inappropriate personal disclosure by the therapist
- Excessively intrusive questioning of a client
- Avoiding referral for fear of exposing therapist incompetence
- Failing to avoid or end a helping relationship, when the therapist lacks competence or objectivity
- Any sexual contact with the client

Sexual Harassment

Counselors are ethically responsible for being aware of the boundaries of behaviors that represent sexual harassment. The counselor is never to intrude on the client through sexual comments, gestures, or propositions.

Fees

The counselor is ethically accountable for making the client aware of any fees for services, and any incidental costs, before the counseling sessions begin. The counselor also ensures that the method for payment has been clearly established and that the client fully understands the counselor's expectation regarding payment. I recommend that the counselor work on a sliding scale to take into consideration financial differences among clients.

Professional Associations

Counselors are expected to have expertise and some professional qualifications to be eligible to join a professional association. The advantage for the client is that the counselor's behavior and professional attitude is held to a recognized standard. For counselors, aligning with a professional organization having a code of ethics provides a standard that will be useful if a question is ever raised about the counselor's professional conduct. It also provides an avenue for consultation, for obtaining malpractice insurance, and for professional development.

Record Keeping

The counseling relationship includes all notes, tapes, videos, assessment instruments, and the like that are related to sessions; they are a confidential part of the counselor/client relationship and should be kept separate from official institution or agency records. (They *may* be subpoenaed by the courts.) Thus it is imperative that all records be as objective as possible and that the counselor refrain from making assumptions or judgments of a subjective nature while note taking.

Consultation

In the counseling relationship, the boundary as to consulting other professionals to assist in the client's care must be made clear to the client. As professionals, we must ensure that we inform the client of our right to consult others and, at the same time, be aware of any conflict of interest that may ensue as a result of the consultation. See the ethical decisionmaking model in an upcoming section of this chapter.

Termination of Counseling

The ethical counselor prepares the client for the ending of the therapeutic relationship before the final session. In other words, the counselor must include termination strategies in the planning process.

Malpractice

In today's world, the word *malpractice* should be engrained in all counselors' heads —not as a negative, but as a positive motivation—so they will be effective and accountable to their clients. Counselors who are clear about their boundaries, stay within their counseling limitations, consult and refer when in question, and follow their code of ethics are being preventive counselors. Those who do all of the above, as well as keep accurate records, and stay up to date with client treatment plans and updates, will be less likely to be burdened with a malpractice suit. If you do all of the above, you will be in a position to prove your *competency* (show you performed to the expected professional standard and were not negligent) in the event you are ever faced with a malpractice suit. This will allow the parties evaluating you to make a judgment in your favor.

A MODEL FOR ETHICAL DECISIONMAKING

From the outset this text has encouraged the human services counselor to obtain new information, explore it, and develop and take responsibility for a personal counseling orientation. Developing an *ethical decisionmaking model* is much the same. I recommend you become knowledgable about policy and procedure, regional law, national law, ethical codes, ethical decisionmaking models, and personal ethics, and develop a model that integrates your own values with those of the community in which you work.

The following is an adaptation of Schulz's integrated ethical decisionmaking model, which we believe is eminently applicable to the human services counselor (Schulz, 1994):

Step 1. Identify the ethical issues of a particular situation.

Step 2. Examine the ethical principles that are important to the situation, including:

- Respecting the sanctity of life
- Not willfully harming others
- Keeping promises
- Caring responsibly

- Being responsible to society
- Respecting peoples' right to determine their fate.

Step 3. Choose the most important principle and begin the action process:

- Generating alternatives and examining the risks and benefits of each alternative
- Securing additional information and/or consulting with colleagues
- Examining the probable outcomes of various courses of action

Step 4. Move from cognitive processes to exploring emotions through the use of emotional decisionmaking techniques:

- Exploring time alone to allow your emotions to integrate with the ethical dilemma being faced
- Deliberating (sleeping on it)
- Projecting the ethical decision into the future and thinking about the various outcomes

Step 5. This last component is the action phase. The counselor needs to develop a plan of action, put it in place, and be prepared to take responsibility and to correct any negative consequences that may arise.

CONFIDENTIALITY

Generally, counselors agree on the value of confidentiality in the client/counselor relationship. However, it is not an absolute. While it is professionally prudent to adhere to confidentiality, counselors do not have privileged communication like pastors, physicians, or lawyers. The Canadian Guidance and Counseling Assocation (CGCA) Guidelines for Behavior (1989) state clearly that they are meant only as a guide. When a question arises about breaking confidentiality, the ethical decision-making model may be valuable in assisting a counselor to make the decision.

Limits of Confidentiality

Some information cannot be held in confidence, from both legal and ethical perspectives. For instance, if a client disclosed that she was sexually involved with a previous counselor, you would be ethically obligated to report the situation. If a 12-year-old client disclosed information about being abused, you would be legally obligated to report it.

Corey (1996) outlines as follows the general circumstances under which client/counselor confidentiality legally *must* be broken:

- When the client poses a threat to self or others
- When the counselor believes a minor client is the victim of incest, rape, child abuse, or other crime
- When the counselor deems the client needs hospitalization
- When a client requests that records be released to a third party
- When the information is subpoenaed in a court action

The Duty to Warn and Protect

It is unethical to impart to the client that all communication will remain in strict confidence. The counselor has an ethical responsibility to explain to the client the situations under which confidentiality would be breached. When the counselor perceives that the client's condition puts the client or others in imminent danger, the counselor has an ethical responsibility to take action. Once again, we recommend that the counselor utilize the ethical decisionmaking model to ensure these actions are prudent.

Maintaining Confidentiality Boundaries

In this information age it is well to address some points to remember when contacting clients through various media (answering machine, voice mail, e-mail, fax):

- Letters should be plainly marked *confidential* on the outside envelope
- There should be *no return address* on the envelope
- *First names only* should be used when addressing individuals, in written or verbal communications
- *Never* leave a message on a client's *answering machine* unless you have the client's written consent to do so
- *Do not send information via facsimile,* especially consent forms (this will render the consent void)

BASIC RIGHTS OF CLIENTS

By educating clients about their rights and responsibilities, you not only empower them, you decrease the possibility of the client's initiating legal action against you. By incorporating information regarding basic rights into your introductory meeting agenda, you can foster the development of a trusting, therapeutic relationship.

Right to Informed Consent

To avoid ethical violations, Schulz (1994) suggests implementation of Margolin's "Three Rs of Counseling": rapport with the client, reasonable behavior, and rights and duties of informed consent. Explaining the following to your client provides a framework for informed consent:

1. All techniques and counseling procedures to be used, including rationale and purpose.
2. The counselor's role and professional qualifications.
3. Discomforts or risks reasonably to be expected during counseling.
4. Benefits the client can expect from counseling.
5. Alternatives to counseling treatment, along with an explanation of their risks and benefits.
6. Assurance that questions about what the counselor is doing will be answered at any time.

7. Assurance that the client can stop counseling and withdraw consent for counseling at any time. (Schulz, 1994)

Rights of Minors

In some locations adolescents are enabled to obtain counseling about birth control, abortion, substance abuse, child abuse, and other crises, without parental consent. It will be necessary to check your region's health care policy on these issues.

Right to Referral

The client has an express right to request that a second opinion be obtained regarding treatment or course of action. As ethical counselors, we have an obligation to help the client get a second opinion. This is based on the expectation that the counselor will have knowledge of appropriate referrals and services.

As a counselor, you also have the ethical responsibility to refer a client whose needs/issues are beyond your area of expertise. Once you have referred a client, you are expected to maintain a position of support until the client is able to meet with the new counselor.

Summary

This chapter is only a checklist or quick reference. It is by no means a complete outline of ethical standards and guidelines. I strongly encourage you to study your local ethical codes and become knowledgeable, to protect yourself and your client.

It is the responsibility of people working in human services to espouse the highest standards of professional conduct and ethics. This means we cannot knowingly allow a colleague to breach ethics without confronting the breach; and, if the colleague is unwilling to address it, we are ethically responsible to report the breach to the appropriate professional organization. We have an obligation to protect the client's well-being, *always,* even at the risk of dissolving professional relationships. Remember: use your ethical decisionmaking model, and *be true to your character and your profession!*

References

Addiction Intervention Association (1996). President's message. *The Beacon* [newsletter] 7 (June/July/Aug.):1–2.

Canadian Guidance and Counseling Association (1989). *Guidelines for ethical behaviour.* Ottawa: Author.

Corey, G. (1995). *Theory and practice of group counseling* (4th ed.). Pacific Grove, CA: Brooks/Cole.

Corey, G. (1996). *Theory and practice of counseling and psychotherapy* (5th ed.). Pacific Grove, CA: Brooks/Cole.

Corey, G., Corey, M., Callanan, P. (1994). *Issues and ethics in the helping professions* (4th ed.). Pacific Grove, CA: Brooks/Cole.

Covey, S. (1989). *The 7 habits of highly effective people: Powerful lessons in personal change.* New York: Simon & Schuster.

Gilliland B. E., James, R. K. (1996). *Crisis interventions and strategies* (2nd. ed.). Pacific Grove, CA: Brooks/Cole.

Fuerst, M. (1995). *Disclosure of third party records: Another clash of the Titans?* Paper presented at the meeting of the Federation of Law Societies of Canada and The Law Society of Newfoundland, St. John's, July.

Schulz, W. (1994). *Counselling ethics casebook.* Ottawa: Canadian Guidance and Counseling Association.

Chapter 3

Checklist on Multiculturalism

DEFINING MULTICULTURALISM IN COUNSELING

The face of society is rapidly and continually changing and counselors are responsible for choosing the path of multiculturalism, reports D. W. Sue (1992). For counseling to be effective, the practitioner must take into consideration the culture of the client. The counselor will benefit from understanding multiculturalism, which involves an an appreciation for the diversity that exists in society with regard to ethnicity, race, gender, class, religion, and lifestyle. Pedersen (1994) suggests that a multicultural perspective strives to provide a framework that recognizes the complex diversity of a pluralistic society; further, that shared concerns link all people regardless of their differences.

It is essential to become culturally aware and to educate yourself about the mix of cultures in your practice. First, as human services counselors we must learn about who we are, become comfortable with our own beliefs, and work out our own prejudices. Only then can we address each new client fully and openly. There are many exciting ideas to be learned from the people who now call North America home. Culturally competent counselors also recognize that differences exist within a culture and avoid stereotyping.

Many ethnic and minority groups underuse health and social service agencies, and one potential explanation for this is that mental health providers have failed to confront multicultural issues. Sue and Sue (1990) contend that the reason for underuse and early termination of service is the biased character of the services themselves: they are frequently antagonistic or inappropriate to the life experiences of the culturally different client, they lack sensitivity and understanding, and they are oppressive and discriminating toward minority clients. If, as a practitioner, you anticipate a variety of clients, it is important to maintain a high level of cultural awareness and modify your strategies to meet the special needs of diverse populations. As professional counselors we must learn about the client's values, interpretations, and beliefs before we can be of any use in helping change to occur. In illustrating the necessity of cultural sensitivity, Sue and Sue (1990) state:

> As mental health professionals, we have a personal and professional responsibility to (a) confront, become aware of, and take action in dealing with our biases, stereotypes, values, and assumptions about human behavior; (b) become aware of the culturally different client's world view, values, biases, and assumptions about human behavior; and (c) develop appropriate help-giving practices, intervention strategies, and structures that take into account the historical, cultural, and environmental experiences/influences of the culturally different client.

Because it is unreasonable and unrealistic to expect that counselors have thorough knowledge of all cultures, it becomes imperative that you (the counselor) have a broad acquaintance with the general principles of various cultures. As you gain experience counseling clients from cultures other than your own, your awareness of multiculturalism will grow. Where you still lack understanding of a particular culture, you can show respect through asking courteous questions. There is no need to feel embarrassed or afraid; generally clients will be pleased that you care enough to learn about their culture.

TERMS RELATED TO MULTICULTURALISM

As you begin to gain awareness of multicultural issues, there are some terms that will become familiar. Some of these are:

- Acculturation—the process of social adaptation of a group or individual.
- Acculturate—to cause acculturation.
- Anomie—a personal state of uncertainty due to a lack of purpose or ideals. A person who can no longer associate with any specific ethnicity may experience anomie.
- Culture—the social and religious structures and intellectual and artistic manifestations that characterize a society.
- Ethnic—descriptive of large groups of people whose unity rests on racial, linguistic, religious, or cultural ties.
- Ethnocentrism—the tendency to regard one's own group/culture as intrinsically superior to all others.
- Ethos—the spirit of a people, a civilization, or a system, as expressed in its culture, institutions, ways of thought, philosophy, and religion.
- Minority—a group distinguished by its religious, political, racial, or other characteristics, from a larger group or society of which it forms a part.
- Multicultural—refers to the complexity of society, which influences service delivery.
- Pluralism—the doctrine that there is more than one universal principle.
- Pluralistic society—a society (i.e., American) that encompasses a variety of ethnic groups and philosophies.

- Race—a distinct group of people, the members of which share certain inherited physical characteristics (i.e., skin color, hair type) and transmit them. (Molecular biologists have recently contended that race may be an invalid way of distinguishing among people.)
- Society—the state of living in organized groups; any number of people associated together geographically, racially, or otherwise; any stage of development of a community.

MULTICULTURALISM AND HUMAN SERVICES

Corey (1995) suggests the following ways to integrate yourself professionally into a multicultural environment:

1. Learn more about the ways that your own cultural background has influenced your thinking and behaving. Become familiar with ways that you may be culturally encapsulated. What are some specific steps you can take to broaden your base of understanding, both of your own culture and of other cultures?

2. Identify your basic assumptions, especially as they apply to diversity in culture, ethnicity, race, gender, class, religion, and lifestyle, and think about how your assumptions are likely to affect your practice as a counselor.

3. Learn to pay attention to the common ground that exists among people of diverse backgrounds. What are some universal concerns shared by all human beings?

4. Realize that it is not necessary to learn everything about the cultural backgrounds of your clients before you begin working with them. Allow them to teach you how you can best serve them.

5. Spend time preparing clients for a successful experience, especially if their values differ from some of the values that are important to you. Teach clients how to adapt their counseling experiences to meet the challenges they face in their everyday lives.

6. Recognize the importance of being flexible in applying the methods you use with clients. Don't get mired in a specific technique if it is not appropriate for a client.

7. Remember that practicing from a multicultural perspective can make your job easier and more rewarding for both you and your clients.

CHARACTERISTICS OF EFFECTIVE MULTICULTURAL COUNSELORS

In 1993, the American Psychological Association, Office of Ethnic Minority Affairs, issued a set of guidelines that included the following:

Effective multicultural counselors have:

1. Knowledge and skills in working with a diverse range of clients. In areas where they do not possess the required knowledge, they seek to fill the void through consultation, supervision, and continued education/training.

2. Awareness of their own cultural background. They are aware of the influence of their personal attitudes, values, beliefs, and biases, and make right and true any prejudices they may have.

3. Accept that ethnicity and culture influence a client's (and their own) behavior.

4. Respect and accept the function and influence of family and community, and their hierarchies.

5. Respect client's religious and spiritual beliefs and values.

6. Ability to aid clients in determining the source of their difficulties, when they arise from the biases and racist attitudes of others.

7. Awareness of social, environmental and political factors, and their impact, when assessing problems and devising interventions.

Over the last decade, counseling and psychotherapy have made exciting advances. The challenge in this decade is to relate this advanced effectiveness of therapy to a varied ethnic and cultural clientele (Ivey, Ivey, and Simek-Morgan, 1993).

Summary

Whether you are working as a counselor in the United States or in Canada, it is imperative to acquire the knowledge, skills, and attitude that are congruent and effective in a pluralistic society. It is important to note that the majority of counseling orientations presented in this text were created, and subsequently practiced, in a predominately white, Anglo-Saxon culture. Thus it is even more important that we become sensitive to the backgrounds of our clients and choose the components of our counseling orientation that are appropriate, prudent, and ennobling. True eclecticism is, in my opinion, not just techniques; it is also awareness, education, and empathy that different cultures require different approaches.

References

Corey, G. (1995). *Theory and practice of group counseling* (4th ed.). Pacific Grove, CA: Brooks/Cole.

Corey, G. (1996). *Theory and practice of counseling and psychotherapy* (5th ed.). Pacific Grove, CA: Brooks/Cole.

Ivey, A. E., Ivey, M. B., Simek-Morgan, L. (1993). *Counseling and psychotherapy: A multicultural perspective* (3rd ed.). Boston, MA: Allyn & Bacon.

Pedersen, P. (1994). *A handbook for developing multicultural awareness* (2nd ed.). Alexandria, VA: American Counseling Association.

Sue, D. W., Arredondo, P., McDavis, R. J. (1992). Multicultural counseling competencies and standards: A call to the profession. *Journal of Counseling and Development 70*:477–84.

Sue, D. W., Ivey, A. E., Pedersen, P. B. (1996). *A theory of multicultural counseling and therapy*. Pacific Grove, CA: Brooks/Cole.

Sue, D. W., Sue, D. (1990). *Counseling the culturally different: Theory and practice* (2nd ed.). New York: Wiley.

The New Lexicon Webster's Encyclopedic Dictionary of the English Language, Canadian Edition. (1988). New York: Lexicon Publications, Inc.

Chapter 4

Professional Assessment for Client Readiness

In the field of counseling, one of the biggest frustrations occurs when the process of counseling does not appear to be working. Many young counselors perceive that because the client is not changing they are not doing a good job. Of course, the counselors may be unaware that they are not doing a good job, perhaps because they have not clarified their own counseling orientation. However, the majority of the time, I find a lack of results signals that the client is not ready to change.

To address this issue, let us consider a frame of reference for assessing the client's readiness for change, along with a model to help move the client through change. Let us say that change has occurred when clients learn new behaviors that help them believe they are in better control of their lives. Prochaska, DiClemente, and Norcross (1992) have developed a six-stage model to assist counselors in determining the readiness of a client and to use as a guide in assisting the client to obtain a healthier and happier direction (Figure 4–1).

In beginning counseling, two common sources of frustration are (1) the client refuses to believe there are any issues requiring change, or (2) the client shows a lack of motivation to change. Prochaska and colleagues (1992) call this the *precontemplation stage.* I have defined it as the *unwillingness to change stage,* where the client is neither psychologically ready nor motivated to address the presenting issue, or has had insufficient time to process the information and come to see the value in a new behavior.

Once the client starts to have dialogue and show willingness for change, Prochaska and colleagues (1992) refer to this as the *contemplation stage.* I have named it the *exploration of change stage,* where the client starts to make inquiries and begins to explore the what-ifs and the potential, the effort required, and the new skills, knowledge, and attitude needed to acquire, to have the potential to achieve the desired change.

Figure 4-1.

Assessment for change (adapted from Prochaska et al., 1992)

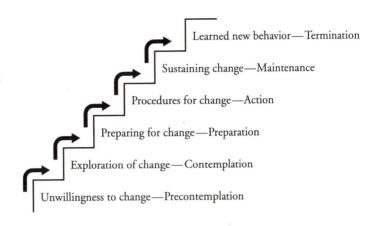

Learned new behavior—Termination

Sustaining change—Maintenance

Procedures for change—Action

Preparing for change—Preparation

Exploration of change—Contemplation

Unwillingness to change—Precontemplation

The point where the client is concentrating and focusing on making change, the authors of this model call the *preparation stage*. I define this stage as *preparing for change stage*. This is where the professional's main function is to act as a resource, and as a teacher to assist the client to develop the foundation skills necessary to be successful in the process of learning new behavior.

When the client is finally ready to initiate change, Prochaska and colleagues (1992) define this as the *action stage*. I call this the *procedures for change stage*. Here, through the counselor's individual eclectic orientation and case plan management, the emphasis is on the client's active participation in exploring and practising the new behaviors that will lead to the desired changes.

The importance of continuing the new behaviors and reinforcing is what the authors of the model define as the *maintenance stage*. I refer to this as *sustaining the change stage*. In the process of learning new behaviors, there is always the potential for minor lapses or relapse. We do not teach the client to expect failure; however, as counselors we cannot be surprised or discouraged when clients slip back, because mistakes are a part of the process of learning (e.g., we all learned how to ride a bicycle, despite falling off, because we got back on). This point cannot be over-emphasized. We must not ever prejudice a client who is learning new behaviors. In concluding this point, we chose the word *sustaining* because of its correlation between repeating behaviors and time—for it takes time, commitment, dedication, and effort for a new behavior to become ingrained and automatic (e.g., passive aggressive behavior changes to assertive behavior).

The final stage of change, according to Prochaska and colleagues (1992), is the *termination stage*. This is when the client shows no interest in old behaviors and is satisfied to choose new ones. I define this as *learned new behaviors stage*. I have chosen to take this position because, as you can see in the cycle of assessment for change of Figure 4-2, the process of change happens both in the larger picture, the client's total happiness (gestalt), and on the smaller fronts, addressing and learning new behaviors to attend to the different issues. This point may be better explained through a brief client profile.

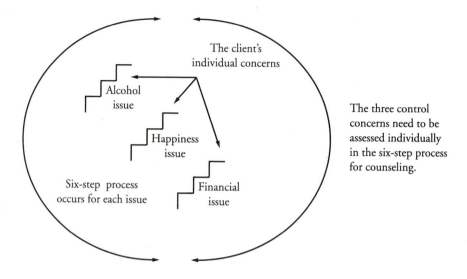

Figure 4–2.
Cycle of
assessment
for change:
Case of Bob.

The client's
individual concerns

Alcohol
issue

Happiness
issue

Financial
issue

Six-step process
occurs for each issue

The three control
concerns need to be
assessed individually
in the six-step process
for counseling.

A Case History

Bob is an unemployed 55-year-old male who is malnourished and impaired by alcoholism. He has recently been divorced and he has no living family. In our initial assessment, the first issue to be addressed was Bob's chronic addiction to alcohol. I recommended detox and Bob agreed. After detox, Bob voluntarily entered an outpatient substance abuse program to address his addiction. At this stage, Bob was motivated to address all of his presenting concerns. A challenging road lay ahead as he looked at issues of grief, loneliness, financial insecurity, and lack of an emotional support system. Many new behaviors will have to be implemented to arrive at the overall goal of personal happiness and fulfilment. Figure 4–2 shows the six-step process occurring continually throughout the entire counseling process.

Acknowledgment

The author is indebted to James Prochaska, Carlo DiClemente, and John Norcross for much of the material included in this chapter. The reader is referred to their *Systems of Psychotherapy: A Transtheoretical Analysis,* 4th edition (Brooks/Cole, 1998), and Prochaska and DiClemente's *The Transactional Approach: Crossing Traditional Boundaries of Therapy* (Dow Jones–Irwin, 1984) for works in which their ideas are more completely delineated.

Summary

In conclusion, this chapter presents a paradigm for assessing if the client is ready for change and then moving with the client through the process toward the client's desired goals. A formula for change must include the acquisition of new knowledge, skills, and attitudes. What Prochaska and colleagues (1992) have taught us is congruent with the philosophy expressed throughout this book, that the choice of change lies within the client (in other words, the client functions from an *internal locus of control*). As in the case of Bob, when you look at a client you can look at the overall motivation for change as well at the motivation for change in each specific concern area.

References

Prochaska, J. O., DiClemente, C. C., Norcross, J. (1992). In search of how people change: Application to addictive behaviors. *American Psychologist 47*:1102–14.

Chapter 5

Assessment and Treatment Planning

No matter what counseling orientation you have chosen, you will begin by assessing your client's needs and then develop a treatment plan. Because all humans are different, each treatment plan is unique in addressing the specific needs of the individual client.

All treatment plans have beginning, middle, and ending stages. When you develop a treatment plan, you need to be aware that *the ultimate goal is termination.* At termination the client leaves counseling with new skills, independence, and the ability to self-manage. It is important to talk about termination early in the relationship. This assures clients that you believe they have the potential to make all the changes needed to achieve their desired outcomes.

Your treatment plan is not window dressing. It is carefully crafted to help the client obtain both short- and long-term goals. In addition to helping the client, a treatment plan provides criteria by which your particular agency can measure quality of care. It also provides proof of treatment for third-party reimbursement. A treatment plan is carried out in conjunction with the client and must be both flexible and client-driven. It allows the client to notice and evaluate progress toward specific goals at any point in the counseling process. A good treatment plan is designed like a series of building blocks, so that the client is taking small steps to the ultimate goals of self-containment, self-control, and termination.

One way to develop a strong treatment plan is to complete a structured assessment interview designed to identify the client's individual needs. Such an assessment tool—the Human Services Counseling Guide—is presented at the conclusion of this chapter. The assessment process is designed to ascertain what the client believes will contribute to greater quality of life.

GOALS OF ASSESSMENT

As you are conducting an assessment interview, keep in mind the following goals:

1. To gather consistent and comprehensive information (best done through a structured format).
2. To identify and define a person's major strengths.
3. To identify the problems that are bringing the client to counseling.
4. To bring a degree of order by prioritizing problems.
5. To teach the inadequacy of the quick fix.
6. To clarify diagnostic uncertainty.
7. To measure cognitive functioning.
8. To differentiate treatment assignments.
9. To develop rapport with the client and create a healthy working environment.
10. To focus on the therapeutic intervention.

And, for those who are abusing substances:

1. To provide an opportunity for the substance-abusing client to associate problems with substance use.
2. To educate about the effects of alcohol and/or drugs, if appropriate.
3. To give clear and unambiguous information regarding addiction and treatment.

One of the secondary benefits of a treatment plan is protection of both the counselor and the client. A treatment plan is a document explaining exactly what you are working on with the client. By contrast, your counseling orientation consists of the theories and techniques you will be using with the client when carrying out the treatment plan. For example:

Client: Johnny Jones
Primary concern: Low self-esteem
Long-term goal: Increased self-esteem **Short-term goal:** To build self esteem by following the assigned interventions.

Interventions/resources/referrals:
1. Individual counseling
2. Self-esteem readings (bibliotherapy)
3. Exercise and diet
4. Daily journal writing and reflections to evaluate self-esteem growth

---

Intervention number 1, *individual counseling*, is where you depend upon your counseling orientation to address self-esteem issues. For example:

Intervention:	**Theory:**	**Techniques:**
Individual counseling	Cognitive, rational-emotive therapy	Teach ABC

As you can see in this brief example, the client's individual treatment plan addresses specific issues and incorporates the treatments and techniques that are part of your counseling orientation. However, your orientation is not merely a theory or technique in action; it actually represents your philosophy throughout the entire treatment process. For example, consider Glasser's *setting the environment,* and the element of *rapport* (Chapter 15): Your focus will remain on assisting the client to feel safe, comfortable, and relaxed no matter which part of the treatment plan you are working on.

Your counseling orientation can be specific as well as abstract. The important thing is that, in formulating any treatment plan, you need to know what your orientation is. Should your treatment of a client ever be challenged, your documented counseling orientation is empirical evidence that what you are doing is based in the theory and standard techniques used in professional counseling. As mentioned earlier, a treatment plan is for the protection of both the client and the counselor. Both a treatment plan and a counseling orientation provide you with a road map of where you and going and what your are helping the client to achieve.

The importance of being well prepared can be emphasized using what I call the four T's of counseling. They are:

1. Theory

2. Techniques

3. Treatment plans (assessment and documentation)

4. Termination

A good counselor will have all four T's well thought out. Remember, as you move on to develop your individual counseling orientation, that every counselor has an individual style and every client needs a unique treatment plan. Treatment planning is a fundamental skill that can be learned through supervision or by taking a course in it. There are also many resources on the market that are designed to help in treatment plan design, one of which is "Therascribe," which was written by Jorgsma, Peterson, and McInnis. Their book, *The Complete Psychotherapy Treatment Planner* (1995), explains this basic model and paradigm. The more you learn about treatment planning, the better; it is an invaluable part of any counselor's orientation.

What follows, and concludes this chapter, is an instrument I designed to help my new counseling students do an assessment and develop a treatment plan. The Human Services Counseling Guide (HSCG) is intended to be used by the counselor with the client from the first interview to the end of the treatment contract. The HSCG provides a structured assessment, treatment planning, organization of resources (e.g, AA), case management, case evaluation, progress notes, and a consent form.

Like all counseling instruments, this tool is no substitute for the counselor's clinical judgement, supervision, or consultation. You may not want to use this instrument in its entirety. The purpose of the HCSG is to act as a guide in working with the client and to facilitate the change process.

PRE-ASSESSMENT CHECKLIST

Before any assessment takes place, I recommend that the counselor consider the following:

1. *Screening.* Clients should be screened to ensure that they meet the criteria of the counselor's agency or practice. If the client does not meet the criteria, I believe it is important to offer a few referral options.

2. *Intake.* The focus of intake is to start the process by building rapport and collecting general information (see items 2.0 to 2.3).

Here are some general suggestions for introducing the client to the assessment process:

- Introduce yourself: *Hello, I'm (title)(name). And you are_____?*
- Explain what an assessment is for, or ask what they are expecting.
- Explain what your role is, or ask the client to define what your job is, and then clarify.
- Explain the concept of choice, and that the counselor cannot make or fix a person; the work is to help the client through areas of concern.
- Explain any orientation information the client may need to know.
- Explain how long the interview will be, and that a break for any reason (for example, the restroom) is OK.
- Explain that the assessment process is ongoing, and that all decisions are made in a collaborative manner. Also, that the client's goals can change or be fine-tuned at any time.
- Explain that this is a helping process, and at any time (unless court ordered) the client can terminate the process.
- Explain, if applicable, that you have already read any reports; however, the emphasis for today is on what the client has to say and wants to do.
- Explain your qualifications.

Summary

As a human services counselor, regardless of your orientation, you will need a tool to do an assessment and develop a treatment plan. Most agencies will have their own format for you to follow. If you are in private practice, you will need to develop your own assessment and treatment planning style, much the same as you develop your own counseling orientation.

Reference

Jorgsma, A. et al. (1995). *The complete psychotherapy treatment planner.* New York: Wiley.

Human Services Counseling Guide

HSCG—Clinical Assessment

(Sections 1.0 –15.0)
Permission to photocopy granted by the author.
Initial and date each page when finished.

1.0 Counselor Administrative Checklist

1.1 Counselor name: _____

1.2 Date of initial contact with client: _____

1.3 Date of this assessment: _____

1.4 Referral source:_____

New client: Yes ☐   No ☐   Last treatment date:_____

1.5 Explain consent: Yes ☐   No ☐   (See Addendum A for consent form.)

If no, explain why _____

1.6 Explain confidentiality: Yes ☐   No ☐

If no, explain why _____

_____

1.7 Discuss financial coverage: Yes ☐   No ☐   Agreed fee:_____

Payment method:_____

If no, explain why _____

_____

Initial and date page when completed _____

Basic Client Intake Information

Client's file number: _____

2.0 Personal identifying information

Last name:_____ First name:_____ Initial: _____

Maiden name (if applicable): _____

Alias (if applicable): _____

Date of birth: _____ Age: _____Yrs. _____Mos.

Citizenship: _____

2.1 Address

Civic address: _____

_____

_____

_____

Mailing address: _____

_____

_____

2.2 Telephone

Home: _____ Business: _____

Leave message: Yes ☐   No ☐

Pager: _____ Cellular: _____ Voice mail: _____

2.3 Living arrangements (have the client rate)

Poor ☐   Fair ☐   Good ☐   Excellent ☐

Comments: _____

_____

_____

Initial and date page when completed _____

2.4 Education

High school:_____

Technical /College /University: _____

Grade point average:_____

Future educational plans: _____

Have you ever been diagnosed with any type of learning disability? Yes ☐   No ☐

If yes: (a) When? (b) By whom? (c) Any medication? _____

_____

_____

What are your present learning strategies? _____

_____

_____

2.5 Present employment\last employer: _____

2.6 Basic multicultural information

Gender: Male ☐   Female ☐

Religion: _____ Preferred language: _____

Race: _____ Sexuality: _____

Cultural Identity: _____

Is there anything you could teach me right now about your culture that may help

facilitate this interview? _____

_____

_____

2.7 Spiritual Information

Religion: _____

Are you a spiritual person? Yes ☐   No ☐

Briefly explain your position: _____

_____

Initial and date page when completed _____

2.8 Legal Status

Do you have a criminal record? Yes ☐   No ☐

If yes: What is your current legal status? _____

_____

What is your criminal record for the last ten years? _____

_____

Have you ever been incarcerated? Yes ☐   No ☐

If yes: When, where, and for how long? _____

_____

3.0 Reason for Assessment

What is your major reason for seeking counseling (self or external, e.g., court)? _____

_____

_____

3.1 How long have you felt like this? _____

_____

3.2 To date, what have you tried to do in an attempt to feel better? _____

_____

3.3 Have you tried counseling for this concern? _____

3.4 Have you ever tried counseling before? Yes ☐   No ☐

If yes: When, with whom, and what did you do in counseling? _____

_____

3.5 What happened that led you to counseling today? When was the last time you felt pain

similar to this? _____

_____

_____

Initial and date page when completed _____

3.6 Do you ever think/see/feel (use the term that best matches the client's modality) a time when you will feel better? What is your perception? _____

_____

3.7 Are you motivated to work to improve your current circumstances? Yes ☐ No ☐

If no: What are you wanting to obtain through counseling? _____

_____

3.8 Pretend that counseling went perfectly for you; what do you want your life to be like after you finish counseling (please do not base your answer on your current circumstances)? _____

_____

3.9 If you were to draw a metaphorical picture that best describes your current situation, what would you draw? (Give the client an example of a metaphor. After the picture is drawn, ask the client to describe what all the different parts represent and what role the counselor takes in this metaphor.)

How You Picture Your life Today
(Draw your current situation metaphor on this page)

Initial and date page when completed _____

4.0   Medical History

How would you rate your physical heath during the past year?   Good ☐   Fair ☐   Poor ☐

4.1   Have you ever been diagnosed with any of the following conditions?

Condition	No	Yes
Sexually transmitted disease (STD)(please specify)	_____	_____
Kidney disease	_____	_____
Diabetes	_____	_____
Tuberculosis	_____	_____
Rheumatic fever	_____	_____
Arthritis	_____	_____
Hepatitis	_____	_____
Epilepsy	_____	_____
Stomach problems	_____	_____
Chronic headaches	_____	_____
High blood pressure	_____	_____
Pancreatitis	_____	_____
Tremors	_____	_____
HIV	_____	_____

Specific injuries: _____

4.2   Have you ever been tested for HIV?   Yes ☐ Last test date: _____   No ☐

4.3   Do you have any specific medical / health problems? _____

_____

4.4   Have you ever undergone a surgical procedure? Complications? Age?

4.5   Psychiatric: Have you previously had counseling? Age? Duration? Reasons?

_____

Type of treatment (medications) ? Your perceived benefit?

(Note: With a psychiatrist, social worker, counselor, psychologist?)

_____

_____

4.6   Obstetrical / gynecological: Have you ever been pregnant? Have you ever had a

miscarriage (spontaneous abortion)? Therapeutic abortion? Postpartum depression?

_____

Initial and date page when completed _____

At what age did your menses begin? At what age did they stop? _____

Have you ever had a Pap smear with abnormal findings? _____

4.7   When did you last have a complete physical examination by your medical doctor? _____

_____

*** If you have had a complete examination within the last six months, please have your physician send me a report. If a complete examination has not been done, please arrange with your physician to have one done, and forward a copy of the result.***

4.8   Do you suffer from environmental illnesses? Yes ☐   No ☐

If yes: Please explain: _____

4.9   Do you practice safe sex ? Yes ☐ How? _____   No ☐

4.91 Physical fitness information

Are you an active person?	Yes ☐	No ☐
Do you exercise?	Yes ☐	No ☐
Do you monitor your diet?	Yes ☐	No ☐
Do you monitor your sleep?	Yes ☐	No ☐
Are you a physically fit person?	Yes ☐	No ☐

Assessor's assessment of client's general health: _____

_____

_____

_____

5.0   Family History and Relationships

Draw a genogram with client.

(Note: See the family systems chapter for information on drawing a genogram.)

5.1   Describe your parents' personalities, and your relationship with them.

Father: _____

5.2   Mother: _____

5.3   Describe your siblings' personalities, and your relationships with them: _____

_____

_____

Initial and date page when completed _____

5.4 If you have a step family, describe their personalities, and your relationships with them.

_____

_____

5.5 If you are adopted, do you know your biological parents? Yes ☐   No ☐

If yes: Do you have contact with them? How is the relationship? _____

_____

If no: Have you ever wanted to find your natural parents ? _____

_____

5.6 Describe (as if you were an author) all of the different roles of the people in your family

(e.g., the caretaker, the comedian, the victim, the unwanted, the outsider).

Name   Character role

_____

_____

_____

_____

_____

_____

5.7 Has anyone in your immediate or extended family ever been incarcerated? Hospitalized

for psychological reasons? Attempted or committed suicide? _____

_____

5.8 Has anyone in your immediate or extended family had a drug/alcohol problem or mental

health concern? Who? Issue?_____

_____

5.9 As you look back, how would you describe your childhood? Poor ☐   Fair ☐   Good ☐

5.91 As you were growing up, was personal safety ever an issue? Yes ☐   No ☐

If yes: Describe _____

_____

Initial and date page when completed _____

5.92 If someone in your family broke a "rule," how was it addressed? _____

_____

_____

5.93 At what age were you first left alone? _____

5.94 A famous psychiatrist developed what he calls the "magic wand technique." Suppose you possessed this magic wand and could change anything about your childhood that you wanted changed. What would it be? _____

_____

_____

6.0 Financial Information

How would you rate your current financial situation? Poor ☐   Fair ☐   Good ☐

6.1 Describe your perception of your debt load:

Unmanageable ☐   Manageable ☐   Not a concern ☐

6.2 Do you worry daily about basic survival needs such as food, shelter, and/or clothing?

Yes ☐   No ☐

7.0 Social Information

Are you a member of any clubs, groups, teams, or other organizations?

7.1 How do you rate yourself? Outgoing ☐   Loner ☐

How long have you been this way? _____

7.2 Are you a leader or a follower? _____

7.3 In a group, are you a thinker or listener? _____

7.4 Do you have true friends? Who?_____

Last contact?_____

7.5 Do you have a social group of friends that you would call healthy and positive influences?

Yes ☐   No ☐

Initial and date page when completed _____

7.6 When was the last time you really enjoyed yourself ? What were you doing? Who were

you with? _____

_____

7.6 What does healthy and safe fun look like for you? _____

_____

8.0 Work History

How old were you when you first obtained a paying job? _____

8.1 Provide a brief history of your last ten years of work? If you changed jobs, explain why.

_____

_____

8.2 How do you rate yourself as an employee?

Fair worker ☐   Good worker ☐   Highly productive worker ☐

8.3 How do you get along with authority figures (bosses) at work?
Fair working relationship ☐
Good working relationship ☐
Excellent working relationship ☐

Counselor's general comments regarding the client's life history: _____

_____

_____

9.0 Alcohol

Have you ever used alcohol? Yes ☐   No ☐

9.1 Were you or others concerned about your use of alcohol during the last twelve months?
Yes ☐   No ☐

9.2 (If the answer is yes to 9.1, have the client complete your choice of accurate measures to assess for alcohol abuse/alcohol dependence suggested instruments).

Note: Choice of alcoholic beverage: _____

Initial and date page when completed _____

Amount of alcohol consumed during episode: _____

Drinking pattern: _____

Duration of this drinking pattern: _____

	Date administered	*Result*
1. MAST	_____	_____
2. ADS	_____	_____
3. SAQ	_____	_____
4. DSM-IV	_____	_____
5. TII	_____	_____
6. Other	_____	_____

9.3 Client's perception regarding status with regard to alcohol: _____

_____

9.4 Assessor's perception regarding client's status with regard to alcohol: _____

_____

10.0 Drug Use (prescription and non-prescription)

10.1 Have you ever used drugs? Yes ☐   No ☐

10.2 Are you or others concerned about your use of drugs during the last twelve months?
Yes ☐   No ☐

10.3 (If client answers yes to 10.2, have client complete your choice of accurate measures to assess for drug abuse/drug dependency.

Note: Choice of drug(s): Primary: _____

Secondary: _____

Amount of drug consumed during episode: _____

Drug pattern: _____

Duration of this drug pattern: _____

Weekly cost of habit: _____

Initial and date page when completed _____

	Date administered	*Result*
1. DAST	_____	_____
2. SAQ	_____	_____
4. Drug Use History	_____	_____
5. TII	_____	_____

10.4 Client's perception regarding drug use: _____

_____

10.5 Assessor's perception regarding client's status with regard to drug use: _____

_____

11.0 Gambling

11.1 Have you ever gambled? Yes ☐   No ☐

11.2 Were you or others concerned about your gambling during the last twelve months?
Yes ☐   No ☐

11.3 (If client answers yes to 11.2, have the client complete your choice of measures to assess _
for gambling concerns.

Note: Type of gambling: _____

Duration of this gambling pattern: _____

Estimate of weekly gambling costs: _____

	Date administered	*Result*
1. South Oak	_____	_____
2. Gambling 20 Questions	_____	_____
3. DSM-IV	_____	_____
4. TII	_____	_____

11.4 Client's perception regarding gambling: _____

_____

11.5 Assessor's perception regarding client's status as to gambling: _____

_____

Initial and date page when completed _____

12.0 PTSD Screen

    12.1 Have you ever been exposed to a traumatic situation, violence (against self or witnessed), sexual abuse, accident, rape, physical abuse, or terrifying event? Yes ☐   No ☐

        Note: Type of trauma: _____

        How long ago did the trauma occur: _____

    12.2 If yes, screen for PTSD (suggested measures).

	Date administered	Result
1. PTSD Screening Instrument	_____	_____
2. DSM-IV	_____	_____

    12.3 Assessor's perception regarding client's status with regard to PTSD:

13.0 Mental Status

Emotions: *Anxiety, hopelessness, depression, suicidal* and *homicidal ideations.*
Rate your mood on a scale of 0 to10, with 0 meaning life is not worth living and 10 meaning that life is great and you are hopeful about life and your future. Rate yourself for this exact moment.

_____
0   5   10

    13.1 (If the client rates emotion level less than five, complete a Beck's Anxiety/Suicide/Depression/Hopelessness tool (as appropriate) with the client.

Test administered	Date administered	Result
_____	_____	_____
_____	_____	_____
_____	_____	_____
_____	_____	_____

14.0 Cognitive Abilities

    14.1 How would you rate your memory over the last six months? Poor ☐ Fair ☐ Good ☐

    14.2 How would you rate your concentration over the last six months?

        Poor ☐   Fair ☐   Good ☐

Initial and date page when completed _____

14.3 (If the counselor has concerns regarding the cognitive abilities of the client, administer the Trailmaking Tests Part A & B (Reitan, 1986), Digit Symbol Subtest of the Wechsler Adult Intelligence Scale, and Mini-Mental State (Folstein, Folstein, and McHugh,1975).

Cognitive abilities test administered	Date administered	Result
1. _____	_____	_____
2. _____	_____	_____

15.0 Counselor Section
Score your basic impressions in the following areas:

Intelligence	Below average ☐	Average ☐	Above average ☐
Energy level	Below average ☐	Average ☐	Above average ☐
Confidence level	Below average ☐	Average ☐	Above average ☐
Speech level	Below average ☐	Average ☐	Above average ☐
Orientation to time and place	Below average ☐	Average ☐	Above average ☐
Attitude	Below average ☐	Average ☐	Above average ☐
General appearance	Below average ☐	Average ☐	Above average ☐

16.0 Client's Multi Axis Diagnosis (as per the DSM-IV)

Axis I _____

Axis II _____

Axis III _____

Axis IV _____

Axis V _____

(Note: OMIT this section unless you have been specifically trained in how to use the DSM-IV. Also, it is important to note that this section is not to be seen as labelmaking or as a diagnostic determination of the client's potential. The purpose is to provide the counselor with the information to assess if the client is within his or her expertise and to assist in treatment plan design. The HSCG paradigm is that the client is the only one who can make any absolute diagnosis or give a label (e.g., alcoholic).

Initial and date page when completed _____

16.1 Multimodel (Lazarus's BASICID Assessment to screen for potential treatment areas)

Behavior (overt behaviors present): _____

Affect (emotions, moods and feelings): _____

Sensation (five senses): _____

Imagery (dreams, pictures, and memories): _____

Cognition (thoughts, opinions, and self-talk): _____

Interpersonal relationships (interactions with others): _____

Drugs/biology (prescriptions and nonprescription drugs, exercise and nutrition, and other

physiological response): _____

16.1 Choice Theory Needs Assessment (List all the things that are presently meeting the client's basic needs. The area where the client is low may be a potential area of focus for treatment.)

Basic Needs

1. Love\belonging links and connections: _____

2. Self-worth\feeling of accomplishment\power over self : _____

3. Freedom of choice: _____

4. Fun: _____

5. Survival (food, shelter): _____

16.1 Is this case within the limits of your expertise? Yes ☐   No ☐

If no: To whom are you **referring** the case ? Date contacted? Does the client have an

appointment? _____

16.2 Before starting the treatment plan, review with the client an inventory of all of her or his

positive strengths, coping behaviors, and environmental supports: _____

_____

Initial and date page when completed _____

Treatment Planning and Case Management

17.0 Biopsychosocial Recognition of Presenting Concerns

Presenting concerns	Biological signs/symptoms	Psychological signs/symptoms	Social signs/symptoms
Primary			
Secondary			
Others			

17.1 Motivation for Treatment

The client is in (circle)   Precontemplation   Contemplation   Preparation

Comments about client's motivation for change: _____

_____

17.2 Addiction Relapse Prevention Plan, for treatment plan if necessary
(Note triggers to avoid and stay-safe behaviors and contacts.)

(a) List hierarchy of triggers to avoid: _____

_____

(b) Plan of positive behaviors and contacts
(*Note:* Have plan at all times so it can act as a survival tool or help provide safe thinking when thinking is potentially dangerous to health.)

1. _____   2. _____   3. _____

4. _____   5. _____   6. _____

7. _____   8. _____   9. _____

Comments: _____

_____

17.3 Abuse stay-safe plan / crisis intervention used (e.g., suicide intervention) if necessary:

_____

_____

Initial and date page when completed _____

18.0 Case Management and Treatment Planning #1

Concern Area	Long-Term Objectives	Short-Term Objectives	Interventions/Resources Using/Referrals (Be sure to make clear to referrals their role in this client's continuum of care. It is the role of the case manager to ensure that all parties involved in treatment understand their roles.)
Primary			
Secondary			
Other			

Treatment Plan Contract # 1

I agree to the above treatment plan and counseling contract of _____ sessions to be reviewed at _____ sessions. If after this treatment plan contract I still need assistance, my counselor and I will make a new treatment plan and contract (Appendix B).

Client's signature: _____

Counselor's signature: _____

Date: _____

Initial and date page when completed _____

18.1 Instruments used for assessment and measurements of client's progress in treatment

Instrument	*Pre-Test Date*	*Post-Test Date*	*Follow-Up #1*	*Follow-Up #2*	*Follow-Up #3*
	Score:	Score:	Score:	Score:	Score:
	Score:	Score:	Score:	Score:	Score:
	Score:	Score:	Score:	Score:	Score:
	Score:	Score:	Score:	Score:	Score:
	Score:	Score:	Score:	Score:	Score:

18.2 Results from any referral testing

_____

_____

_____

19.0 Evaluation of treatment plan(s) (*Note:* Complete in conjunction with the client. The client and counselor must agree for this concern area to be dated as goal achieved.)

Concern Areas	*Desired Long-Term Goal Obtained*
1.	Date:
2.	Date:
3.	Date:
4.	Date:
5.	Date:
6.	Date:

19.1 Termination (*Explain why the client has stopped counseling, e.g., has completed treatment, has chosen not to make a commitment*): _____

_____

Closing comments: _____

_____

Initial and date page when completed _____

20.0 Progress notes   No._____   Client _____

Date	Progress Notes	Page _____	Orders and Signature

Initial and date page when completed _____

Addendum A

Consent for Release of Information / Request of Confidential Information

RE: _____ DOB: _____

Having had confidentiality explained to me, I fully understand that the purpose of this form is to

Release \ Request information. I hereby authorize _____

(Name of Organization) to **Release \ Request** the following information to \ from:

Persons / Organization	Yes	No	Initials
1. _____	_____	_____	_____
2. _____	_____	_____	_____
3. _____	_____	_____	_____
4. _____	_____	_____	_____

Area of Disclosure

	Yes	No	Initials
1. Assessment and treatment planning	_____	_____	_____
2. Progress report	_____	_____	_____
3. Legal status	_____	_____	_____
4. Other	_____	_____	_____

Note: I understand no other information will be released without my written consent or a court order or for your safety or the safety of others. I also understand that I can withdraw my consent at any time for any reason, the form must be completed to be valid, and this form is only valid from the time signed for 180 days.

Client's signature:_____ Date:_____

Witness:_____ Date:_____

Initial and date page when completed _____

Chapter 6

Drug, Alcohol, and Gambling Screen (DAGS)

ASSESSING ADDICTION

An important component of any assessment interview is a useful measure to predict potential difficulties with addictions (drugs, alcohol, gambling). For any measure to be useful, it must be (a) user-friendly, (b) quick, and (c) have validity and reliability. Since we found no single measure, my students and I developed one; we call it the Drug, Alcohol, and Gambling Screen (DAGS).

The DAGS is designed to be an effective screen for determining whether further assessment is required as to drugs, alcohol, and gambling. The obvious benefit of the DAGS is that it allows the assessor to confirm or develop a hypothesis of potential addiction concerns while using only one measure.

The questions on the DAGS are based upon the criteria of the *Diagnostic and Statistical Manual of Mental Disorders, Fourth Edition (DSM-IV)*. The DAGS consists of 45 questions, 15 pertaining to each of the three areas of interest (drugs, alcohol, gambling). This screen is scored on a Likert Scale and the score obtained in each of the three domains indicates the severity of use. The DAGS scores indicate potential risk of addiction concerns on the following scale: mild (1–15) , moderate (16–30), medium (31–40), and serious (41–60).

Screening for addictions is a very important part of your role as a human services counselor. The DAGS is not intended to be a labelmaker, but to assist in discovering or ruling out potential concerns. As a human services counselor, you will need a variety of tools to confirm your assessment.

Summary

The DAGS is included in this text to provide a user-friendly measure to score for potential alcohol, drug, and gambling concerns. If you do not have specific training in addiction issues, I recommend that you obtain further training in the field. The DAGS, which follows this summary, is an appropriate tool to assist you in making referral decisions when appropriate.

Reference

American Psychiatric Association. (1994). *Diagnostic and statistical manual of mental disorders* (4th ed.). Washington, DC: Author.

Drug, Alcohol, and Gambling Screen

DAGS

The 45-question DAGS is used in conjunction with the Human Services Counseling Guide.
Permission to photocopy granted by the author.

Note: The questions are to be answered with regard to your present behavior and your behavior over the past 12 months, including today.

1. Please read each question carefully and circle the choice that best suits your current situation.

2. This instrument will take only 15–20 minutes. Please answer all of the questions. If you have a difficult time finding the correct answer for your situation, please choose the most appropriate answer.

3. This screen is to predict the potential risk level for drugs, alcohol, and gambling. The questions regarding drugs pertain only to drugs, not alcohol.

1. Do you feel the need to gamble with increasing amounts of money in order to achieve the desired excitement?
    0-Never   1-Sometimes   2-Often   3-Almost always   4-Always

2. Do you gamble as a way of escaping problems? (work, relationships, family, school)
    0-Never   1-Sometimes   2-Often   3-Almost always   4-Always

3. Have you been able to have one or two drinks and then stop drinking?
    0-Never   1-Sometimes   2-Often   3-Almost always   4-Always

4. Has anyone ever shown concern about your drug use?
    0-Never   1-Sometimes   2-Often   3-Almost always   4-Always

5. Have you missed meals because of prolonged drinking?
    0-Never   1-Sometimes   2-Often   3-Almost always   4-Always

6. Have you disregarded health issues in order to keep drinking?
    0-Never   1-Sometimes   2-Often   3-Almost always   4-Always

7. Has your use of drugs brought you in contact with the legal system?
    0-Never   1-Sometimes   2-Often   3-Almost always   4-Always

8. Have you driven an automobile after having used drugs?
    0-Never   1-Sometimes   2-Often   3-Almost always   4-Always

9. Have you ever borrowed from others (family, friends, bank, loan shark) in order to relieve a desperate financial situation caused by gambling?
    0-Never   1-Sometimes   2-Often   3-Almost always   4-Always

10. Have you disregarded health issues in order to keep using drugs?
    0-Never   1-Sometimes   2-Often   3-Almost always   4-Always

11. Do you lie to family, friends, or others to hide the extent of your involvement with gambling?
    0-Never   1-Sometimes   2-Often   3-Almost always   4-Always

12. Have you repeated unsuccessfully in efforts to control, cut back, or stop gambling?
    0-Never   1-Sometimes   2-Often   3-Almost always   4-Always

13. Do you ever go out drinking and the next day forget what happened the night before?
    0-Never   1-Sometimes   2-Often   3-Almost always   4-Always

14. Do you experience feelings of remorse or guilt as a result of your drug use?
    0-Never   1-Sometimes   2-Often   3-Almost always   4-Always

15. Have you lost interests in activities and friends due to your drug use?
    0-Never   1-Sometimes   2-Often   3-Almost always   4-Always

16. Have you had problems at school or work related to your drug use?
    0-Never   1-Sometimes   2-Often   3-Almost always   4-Always

17. Have you attempted to hide your drinking behaviors (i.e., hiding alcohol, lying to cover up, etc.)?
    0-Never   1-Sometimes   2-Often   3-Almost always   4-Always

18. After losing money gambling, did you feel you must return as soon as possible and win back your losses?
    0-Never   1-Sometimes   2-Often   3-Almost always   4-Always

19. Has drinking caused problems in any of your relationships with family, friends, or significant others?
    0-Never   1-Sometimes   2-Often   3-Almost always   4-Always

20. Do you have difficulty in getting drinking off your mind?
    0-Never   1-Sometimes   2-Often   3-Almost always   4-Always

21. Have you ever sold anything to finance your gambling (family possessions, stocks)?
    0-Never   1-Sometimes   2-Often   3-Almost always   4-Always

22. Have you tried to cut down on your drinking and failed?
    0-Never   1-Sometimes   2-Often   3-Almost always   4-Always

23. Have you ever neglected family obligations because of your drug use?
    0-Never   1-Sometimes   2-Often   3-Almost always   4-Always

24. How often have you made efforts to cut down or quit using drugs, and failed?
    0-Never   1-Sometimes   2-Often   3-Almost always   4-Always

25. Have you gambled until your last dollar was gone?
    0-Never   1-Sometimes   2-Often   3-Almost always   4-Always

26. Have you ever been reluctant to use "gambling money" for normal expenditures?
    0-Never   1-Sometimes   2-Often   3-Almost always   4-Always

27. Do you ever have the urge to celebrate any good fortune by a few hours of gambling?
    0-Never   1-Sometimes   2-Often   3-Almost always   4-Always

28. Have you drunk continuously for twelve hours or more at a time?
    0-Never   1-Sometimes   2-Often   3-Almost always   4-Always

29. How often do you think about cutting down, controlling, or quitting using drugs?
    0-Never   1-Sometimes   2-Often   3-Almost always   4-Always

30. Have you ever used drugs in larger amounts over a longer period than was intended?
    0-Never　　1-Sometimes　　2-Often　　3-Almost always　　4-Always

31. Do you feel that you now use more of a drug to get the same effect that you got when you first were using that drug?
    0-Never　　1-Sometimes　　2-Often　　3-Almost always　　4-Always

32. Do you now use more alcohol to get the same effect that you got when you first started to use alcohol?
    0-Never　　1-Sometimes　　2-Often　　3-Almost always　　4-Always

33. To cover up your drinking, have you ever lied about where you were going?
    0-Never　　1-Sometimes　　2-Often　　3-Almost always　　4-Always

34. Have you ever had difficulty sleeping because of gambling?
    0-Never　　1-Sometimes　　2-Often　　3-Almost always　　4-Always

35. Have you experienced "blackouts" (total loss of memory for any length of time, without passing out)?
    0-Never　　1-Sometimes　　2-Often　　3-Almost always　　4-Always

36. Do you experience feelings of discomfort or anxiety when you are not using drugs?
    0-Never　　1-Sometimes　　2-Often　　3-Almost always　　4-Always

37. Have you ever gambled longer than you planned?
    0-Never　　1-Sometimes　　2-Often　　3-Almost always　　4-Always

38. While drinking, do you continue to drink to excess with no regard to what your responsibilities are?
    0-Never　　1-Sometimes　　2-Often　　3-Almost always　　4-Always

39. Have you ever committed illegal acts (forgery, fraud, theft) in order to finance your gambling?
    0-Never　　1-Sometimes　　2-Often　　3-Almost always　　4-Always

40. Has gambling jeopardized or lost a significant relationship, job, educational, or career opportunity?
    0-Never　　1-Sometimes　　2-Often　　3-Almost always　　4-Always

41. Has anyone ever expressed concern over your drinking behavior?
    0-Never　　1-Sometimes　　2-Often　　3-Almost always　　4-Always

42. Do you consider suicide as a result of your drug use?
    0-Never　　1-Sometimes　　2-Often　　3-Almost always　　4-Always

43. Has the use of drugs affected how well you are able to get along with people in your life?
    0-Never　　1-Sometimes　　2-Often　　3-Almost always　　4-Always

44. Have you ever considered self-destruction as a result of your gambling?
    0-Never　　1-Sometimes　　2-Often　　3-Almost always　　4-Always

45. Do you spend large amounts of time getting or thinking about how to get drugs?
    0-Never　　1-Sometimes　　2-Often　　3-Almost always　　4-Always

The End　　　　Thank You

DAGS Hand Scoring Sheet

Drugs	Alcohol	Gambling
4. _____	3. _____	1. _____
7. _____	5. _____	2. _____
8. _____	6. _____	9. _____
10. _____	13. _____	11. _____
14. _____	17. _____	12. _____
15. _____	19. _____	18. _____
16. _____	20. _____	21. _____
23. _____	22. _____	25. _____
24. _____	28. _____	26. _____
29. _____	32. _____	27. _____
30. _____	33. _____	34. _____
31. _____	35. _____	37. _____
36. _____	38. _____	39. _____
42. _____	41. _____	40. _____
45. _____	43. _____	44. _____

Total: _____   Total: _____   Total: _____

Drug Risk: _____   Alcohol Risk: _____   Gambling Risk: _____

Write Risk Level Below Each Total

No concerns 0–2

Mild concern 3–15

Moderate concern 16–30

Medium concern 31–40

Serious concern 41–60

Chapter 7

Developing Your Integrative Counseling Orientation

The first six chapters of this text introduced some of the basic tools needed by a counselor, regardless of orientation. Now, this chapter explains how to develop your own personal counseling orientation. Many of your choices will grow out of the way you like to address people, because your counseling orientation is simply an expression of you, using counseling theory and techniques that are comfortable for you.

The chapters of Part II will present the theory portion of this text. There you will see visual images called **mind maps** that will assist you in grasping the particular theory presented in each chapter. Later, in the Appendix of this book, you will find a variety of counseling orientations developed by students who have used this "toolbox" as a guide to developing a counseling orientation.

Fundamental to developing your orientation is that you have both basic skills and an understanding of counseling theory. Once these are in place, you just begin—through practice, supervision, and experience—to develop an integrative counseling orientation. It is important to point out that, even after you have developed your orientation, you may need to adapt it to match your client's current needs and style.

MCGREGOR X-Y THEORY

The following continuum is an educational tool based on the McGregor X-Y theory, which I have adapted to help you grasp the various theories more readily as you develop your individual counseling orientation. In 1960, Douglas McGregor defined two opposing styles of leaders or managers. The Theory X leader is autocratic and directive, while the Theory Y leader is democratic and less directive. Theory X and Theory Y each designate a style or approach when applied to the practice of counseling. When I apply this theory to the concept of counseling, the result looks like this:

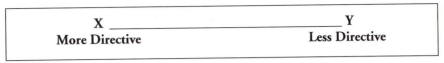

X ——————————————————————— Y		
More Directive		**Less Directive**

Theory X, in the McGregor proposition, makes the assumption that people need strong direction (McGregor, 1960). The counselor at this end of the continuum will use authoritative techniques toward the attainment of certain outcomes; this counselor will actively influence and challenge the client to make decisions and implement certain actions in pursuit of a particular goal. The following are examples of "directive," or Theory X, counseling techniques:

- *Decentering.* The client is encouraged to imagine how a situation or event would look to someone with a different perspective. In doing so, the client's own perspective of that situation or event may be enhanced (Todd and Bohart, 1994).

- *Homework.* The client is give a specific activity or task to carry out prior to the next meeting with a counselor. For example, the depressed client is assigned the daily task of writing down one activity and the thoughts and feelings that accompany it. This information is used at the next meeting to help the client look at thoughts and to introduce the concept of self-monitoring of thoughts (Beck and Weishaar, 1989).

At the opposite end of the continuum is Theory Y, where McGregor posits that people are self-directed and motivated, and will work toward goals they want to achieve (McGregor, 1960). Therefore, the Theory Y counselor *with the same client* will use techniques that are less directive to realize a counseling goal. Following are examples of "less directive," or Theory Y, techniques:

- *Reflecting.* The counselor responds to a client's statement by reflecting or mirroring the client's experience (Todd and Bohart, 1994).

- *Empathy.* The counselor expresses genuine interest in what the client is saying. This interest is not only for the content or meaning but also for the feeling behind what is being said (Rogers and Raskin, 1989).

The marker on the X-Y continuum at the beginning of each of the theory chapters has been adjusted to demonstrate where that theory's particular counseling techniques are located conceptually. The X-Y graphic, therefore, reflects techniques that tend to be more directive (X side of continuum), less directive (Y side of continuum), or gravitate between the two opposite poles. It is important to note that my X-Y continuum does not necessarily indicate how the originators of the theories facilitated counseling and/or psychotherapy. The continuum represents my view of the integration of their techniques with the X-Y concept as they would be used by the human services counselor. Checking out the continuum can give you a starting point for technical development and for formulating your counseling orientation.

DIRECTIVE AND LESS-DIRECTIVE TECHNIQUES

In the past, a common way of describing the difference in counseling orientations has been to designate them as directive or non-directive. Scissons (1993), on the other hand, characterizes this breakdown as outdated and confusing: "The internal processes of all counseling orientations are directive. In each, the counselor had a job to do, although the job differs from orientation to orientation." In other words, counseling is generally directive. However, there are distinct advantages to placing the techniques used by the human services counselor on the X-Y continuum.

In contemplating the X-Y continuum, perhaps the most important thing to consider is yourself. It is the characteristics of your own personality that are most likely to determine where you are comfortable on the continuum and where you will be most effective as a human services counselor. If, for example, your are somewhat reserved, enjoy solitary pursuits, and find that people are always thanking your for being a good listener, you may be most comfortable and congruent carrying out the techniques located near the Y (less-directive) end of the continuum. If you are very outgoing, always take the initiative, and are seen as a real go-getter, you will probable be more comfortable and effective with those techniques that appear toward the X (more-directive) end of the continuum.

YOUR BELIEF SYSTEM

Another important consideration in the choice of a counseling orientation is the belief system of the counselor. All counseling theories propose an *interpretation* of human nature, from which a certain style arises. Ultimately, your beliefs about how people interact and how they are motivated to achieve will determine what counseling orientation you formulate. For example, one tenet of Carl Rogers' person-centered therapy is his actualizing tendency theory, which proposes that "people automatically tend to move in the direction that realizes their full potential"(Rogers and Raskin, 1989). The counselor holding this belief about human nature may be drawn to less-directive techniques on the counseling continuum. Rogers (1965) stated that counselors have a set of attitudes employed in their counseling techniques, and these techniques are consistent with their attitudes. The counselor at the Y end of the continuum believes that individuals are altruistic, assume responsibility, work toward goals, are capable of making their own decisions, and are motivated to change.

In contrast, a counselor on the X end of the continuum may be convinced that clients need clear direction in making decisions, find change difficult, and avoid responsibility. A counselor from the X position is more definitive, scientific, and precise in approaching human nature and may adopt a theory that is similarly exacting. This individual might choose, for example, rational-emotive behavioral therapy, which has cognitive, affective, and behavioral components. In this theory of human nature, people are seen as having a tendency to be irrational, although they do have the potential to be rational (Ellis, 1989).

MIND MAPS

In addition to the X-Y continuum, in each chapter I have included a mind map of techniques. Mind maps are visual illustrations of a variety of concepts (Scheele, 1993). The core concept is located at the middle of the map. Connecting lines radiate outward to supporting concepts, which may themselves branch off to form other subgroups.

Mind maps provide an alternative to the traditional linear outline of educational material. They are considered to be very effective learning tools because they provide a quick reference, promote long-term retention, and are an excellent way to synthesize information (Scheele, 1993). As an educational tool, mind maps permit an understanding of the relationship component at a glance. MacDonald (1994) describes mind maps as a "whole brain visual, interesting version of outlining."

Mind maps are also highly individualized. I encourage you to develop a mind map that illustrates your counseling orientation. This map will have your own stamp of originality, yet still reflect the basic concepts of the counseling theories you have chosen to incorporate into your personal counseling orientation. When creating your mind map, Scheele recommends that you use paper larger than standard business stationery.

I have included a mind map with each chapter of this book. Each design is original and visually unique to the chapter it represents. Please note: *These maps are not full representations of the theories they delineate.* They simply and briefly depict the counseling techniques that have developed from the theory presented in the chapter and are intended to be both relevant and realistically applicable to the field of human services counseling.

UTILIZING PART II OF THIS BOOK

Each chapter in Parts II and III of this text presents a particular theory in the following way.

- An *X-Y continuum* indicates how directive the chapter's techniques tend to be.
- A *mind map* provides a visual overview of the theory and techniques.
- A brief *history* of the theory and its originator follows.
- A description of the *main ideas,* or philosophy, of the theory (e.g., its interpretation of human nature) ensues.
- Following this are a series of *techniques* that are used to put the described theory into practice. *Examples* demonstrate how to implement the technique.

PUTTING IT ALL TOGETHER

Checklist

Armed with some fundamentals and the concept of the mind map, how do you begin the process? Let's begin with a basic checklist for developing your counseling orientation.

- Personal ethics
- Professional ethics
- Multicultural considerations
- Counselor's personality
- Counseling population
- Counselor's employment and role
- Counselor's employment policy and procedures
- Assessment and treatment plans for the population

Having completed the preceding checklist, you are ready to begin formulating your orientation. (If you become confused, see the Appendix for student examples of counseling orientations using mind maps.) The tips that follow may also help to get you started.

Twenty Tips for Getting Started

1. *Rapport.* Remember there is no such thing as resistance. When you think a client is resistant and does not want to talk, focus on building the climate. Remember Dr. Seuss's *Green Eggs and Ham.* If one thing is not working, try another, and never give up. Always show courtesy and a nonjudgmental stance.

2. *Confidentiality and consent.* Verbal or written.

3. *Screening.* Ask yourself: Can I work with this client? This includes intake, plus orientation to any rules or limitations of your organization.

4. *Contracting.* This includes anything from the length of the session to a treatment plan.

5. *Client's expectations from counseling.* What has he or she tried before? What has been heard about counseling, the agency, or you?

6. *Counselor's expectations of counseling.* What you believe are the benefits (or any potential risks) of counseling.

7. *Counselor's qualifications.* Share any limitations that are obvious.

8. *Multicultural recognition.* Be aware of any multicultural differences (e.g., race, religion, gender, age, ethnicity).

9. *Language difference.*
   - *Foreign language.* Native language different from your own.
   - *Regionalisms.* Words or phrases used in particular geographic regions. (Instead of "Hi," in the southeastern United States a person may say "Hey," which can sound rude to people from other regions.)
   - *Colloquialisms.* Conversational words or phrases, such as "getting down to brass tacks"; they are usually used in many regions, and are long-lasting terms.
   - *Slang.* Words or phrases popular for short periods of time, such as "bad" meaning good.

Note: To be able to communicate with clients, we need to learn their (subjective world's) language, and not expect them to change to ours.

10. *Setting the environment.* Ensure that the client is comfortable, with a comfortable chair, soft lighting, and no unpleasant or unwanted sounds or odors (be environmentally friendly, use no strong perfumes), and that the restroom is clearly available.

11. *Stopping the counseling session.* It is important to let clients know they can take a break any time they need to, that the session is meant to be helpful, and that the counselor is working for them. In other words, the client is the true manager, and the counselor is the guide. There may be exceptions to the rule, such as in court-ordered referrals or in certain institutions.

12. *Termination.* In most cases, termination should be brought up in the first session, so that clients know the counselor expects them to get better and be able to move ahead on their own. For example, make a contract about the number of sessions in the treatment plan.

13. *What are you doing that may be disruptive to the client?* Things such as note taking, recording, and videotaping may make clients uncomfortable. The client needs to be informed of the purpose, and be assured about confidentiality, and the client *must give consent.* Warn the client regarding potential interruptions, and avoid these if at all possible.

14. *Be yourself.* Be real, and be true to your values, personal ethics, professional ethics, and the law.

15. *Use humor.* Counseling can be fun, as long as it is never at the client's expense. Be sure the client understands and appreciates your humor. (This is very important!)

16. *Always deal with presenting crisis.* Do not try to move forward until the client is stable.

17. *Focus on your communication.* Be aware of your nonverbal communications. Remember, it is not what you say, it is how you say it. The client's response to your words and body language will tell you if you are making progress.

18. *Continually assess the client's readiness for change.* Is the client ready to move forward?

19. *Know your counseling orientation.* Know well at least one theory, a therapy model, a communications model, some counseling techniques, how to do a treatment plan, where to look for research, and how to evaluate the client's progress. Have a referral system in place.

20. *Believe in people.* Above and beyond everything, always believe people can change, and never give up hope.

Acknowledgments

The author is deeply indebted to Douglas McGregor for much of the material included in this chapter and for the concept of a continuum that is carried throughout this text. The reader is referred to *The Human Side of Enterprise* (McGraw-Hill, 1960) for a more complete explication of McGregor's ideas. In addition, check out Caroline McGregor and Warren G. Bennis, who edited *The Professional Manager* (McGraw-Hill, 1967) and *Leading and Motivation* (with Edgar H. Schein, MIT Press, 1966).

Summary

This chapter's structure reflects my belief that theory and technique are intertwined. Through careful scrutiny of these theories and techniques to come, you will see how it is possible to piece together an eclectic approach based on one or more theories. The techniques you adopt may run the length of the X-Y continuum. My journey to develop my own counseling orientation has been both exhilarating and challenging. I hope your journey will be equally as valuable in your professional development.

References

Beck, A. T., Weishaar, M. (1989). Cognitive therapy. In R. J. Corsini, D. Wedding (eds.), *Current psychotherapies.* Itasca, IL: F. E. Peacock.

Ellis, A. (1989). Rationale-emotive therapy. In R. J. Corsini, D. Wedding (eds.), *Current psychotherapies.* Itasca, IL: F.E. Peacock.

Gladding, S. (1991). *Group work: A counseling specialty.* New York: Macmillan.

MacDonald, J. (1994). *Journey to wholeness.* Ottawa: Health and Welfare Canada, Family Violence Division.

McGregor, D. (1960). *The human side of enterprise.* New York: McGraw-Hill.

Rogers, C. R. (1965). *Client-centered therapy: Its current practice, implications, and theory.* Boston: Houghton Mifflen.

Rogers, C., Raskin, N. (1989). Person-centered therapy. In R. J. Corsini, D. Wedding (eds.), *Current psychotherapies.* Itasca, IL: F. E. Peacock.

Scheele, P. (1993). *The photoreading whole mind system.* Minneapolis: Learning Strategies Corporation.

Scissons, E. H. (1993). *Counseling for results: Principles and practices of helping.* Pacific Grove, CA: Brooks/Cole.

Todd, J., Bohart, A. C. (1994), *Foundations of counseling and clinical psychology.* New York: Harper-Collins.

Part II

Foundations in Theory

Chapter 8

More Directive   Less Directive

Person-Centered Therapy: Carl Rogers

When the counselor perceives and accepts the client as he is, when he lays aside all evaluation and enters into the perceptual frame of reference of the client, he frees the client to explore his life and experiences anew, frees him to perceive in that experience new meanings and goals.

Rogers, 1965a

HISTORY

Carl R. Rogers (1902–1987), the originator of person-centered therapy, was born in Oak Park, Illinois. Rogers earned a history degree, then moved to New York City to study theology. While there, he became interested in psychology. Switching fields, Rogers completed his doctorate in psychology in 1931 and then worked for twelve years as an administrator and child psychologist in Rochester, New York.

The history of person-centered therapy began with Roger's publishing of *Counseling and Psychotherapy* (1942), which announced his thesis that "certain qualities and client conditions (relationship variables) were sufficient to achieve therapeutic change, . . . [which] has substantially influenced both counselor training and the nature of counseling ever since" (Goodyear, 1987). After Rogers moved to California, the term *person-centered approach* began to replace the earlier term *client-centered* (Corsini, 1994). In 1964, Rogers joined the staff of the Western Behavioral Sciences Institute. Then, in 1968, he and his associates formed The Center for Studies of the Person in La Jolla, California, where he worked with both individuals and groups. The influence of his work was recognized in his final years of life when Rogers was nominated for a Nobel Peace Prize.

For more information on person-centered therapy, contact The Center for Studies of the Person, 1125 Torrey Pine Road, La Jolla, CA 92037.

MAIN IDEAS

I have included the person-centered approach in this text because it is the benchmark from which most counselors develop their humanistic underpinnings. Professionals who are beginning their journey to become counselors will find Rogers' work a non-invasive approach with fundamental ideas that are useful in any orientation. In the person-centered approach, the counselor acts as a genuine, caring person who

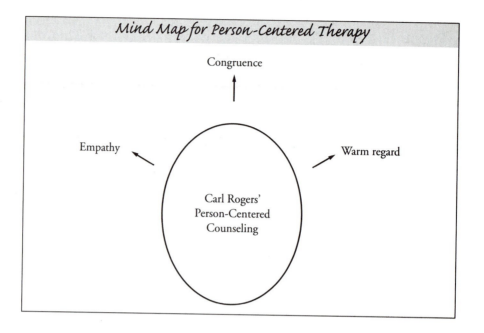

respects the needs and abilities of the client. The Rogerian counselor attempts to create this atmosphere for the client based on the following:

- Rogers believed that a person's inner resources could only be brought out through what Rogers called *basic optimism* (Thorne, 1992). This optimism is the belief that individuals know what they need and how best to obtain it. Rogers felt that, to support a person's inner resources, the counselor needs to get a true understanding of the client's subjective experiences.

- The underlying tenet of the Rogerian approach is the *actualizing tendency of persons,* which is defined as trust in a constructive, directional flow toward the realization of each individual's potential (Rogers and Raskin, 1989). Each human being is thought to possess an innate ability for a self-directed process of growth and fulfillment.

- In person-centered therapy, *warm regard, congruence, and empathy* are defined as the necessary core conditions to facilitate client growth and change. These may also be viewed as specific techniques in the person-centered counseling process. Refer to the techniques section of this chapter for further definition and clarification.

- The counselor, by accepting the world as the client perceives and values it, reduces the resistance of the client and encourages personal growth, giving the client an opportunity to develop a positive sense of self-worth, personal direction, and a capacity to take personal risks (Todd and Bohart, 1994).

- Rogers and Raskin (1989) identify the main principle behind Roger's humanistic philosophy as a personal self-concept based on the idea of locus of evaluation and experiencing. They define *self-concept* as referring specifically to the person's perceptions and feelings about self; this is the primary focus of Rogerian counseling. During the person-centered process, there is a shift from an external to an internal locus of evaluation as the client becomes more self-accepting and regains control of life. The client also moves from an inflexible and rigid stance to one of openness and flexibility. This is called *experiencing.*

- When the core conditions of Rogerian counseling assist the client in having experiences that enhance self-concept, the results lead to what Rogers defines as *self-actualization.* In the self-actualization process, the self-concept of the client experiences a reduction in stress and an increase in self-worth, enabling the client to overcome the concern that originally brought her or him to counseling (Rogers, 1986).

- "According to Rogers, the counselor was not to guide, to reassure or support; he was not to interpret and was not to use an entire armamentarium of what were labeled directive standard techniques . . . it was recommended instead that the therapist stress what were called non-directive techniques" (Saulnier, 1971).

TECHNIQUES

The influence of Carl Rogers' philosophy continues to be felt in many aspects of life, both inside and outside the therapeutic realm, including workshops, organizational (business) development, and concepts of world leadership. His approach can be carried into many orientations as a growth-enhancing therapeutic component.

The following are definitions of core concepts along with examples of person-centered techniques.

1. *Empathy* (core condition). Empathy occurs when the counselor responds to a client's statement with a profound interest in the client's world of meanings and feelings as the client is willing to share this world (Rogers and Raskin, 1989). This empathic understanding is essential to the person-centered counseling process. Rogers (1965a) proposed that, by using empathy, the counselor is also expressing genuine belief in the client's ability to identify and solve personal problems. For example:

   CLIENT: I dont know what to think. First, there was the accident at school; then later I learned my son needs to be hospitalized. To make things worse, my boss at work has been treating me unfairly.

   COUNSELOR: It certainly sounds as if it has been a very challenging week for you.

   CLIENT: Boy, it sure has!

2. *Congruence* (core condition). This refers to the counselor's consistency in thought and behavior; the counselor is to be transparent, authentic, genuine, and honest (Cavanaugh, 1990). Rogers (1965a) notes that counselors cannot act genuine through any form of trickery; they must be truly committed to being

genuine. Rogers defines genuineness as the characteristic of being real and true to oneself (Rogers and Raskin, 1989). The counselor does not show a false front or cover up true feelings; the congruent counselor is honest and natural in relating to the client. Congruence is a basic principle used throughout the counseling process. For example:

CLIENT: I am so angry at my boss for firing me that I want to get revenge.

COUNSELOR: I can see that you're very angry for being dismissed from your job. I will help you all I can to deal with it, but I will not help you to get back at him.

3. *Warm regard* (also called *unconditional positive regard*) (core condition). This is defined as nonjudgmental caring, in which the counselor's positive regard for the client does not change no matter what choices the client makes or what the client does or says. Every action, reaction, or non-action is seen as the valid choice for the client to make (Rogers, 1965a). This is expressed not only through the counselors empathy and congruence but also through an active acceptance of what the client is saying or doing. The counselor responds in a positive and prizing manner, which encourages the client to help himself or herself (Roger and Raskin, 1989). For example:

CLIENT: I get angry, very angry, with her when she wants money.

COUNSELOR: You get angry because of how she spends money.

The following counseling procedures are not all Rogerian. I have chosen to put these procedures in this chapter because I believe all professionals should work from a humanistic perspective. It is of paramount importance that all professionals inform their clients of the issue of confidentiality as well as the limits to confidentiality. As a counselor, you must receive consent from the client to act as his or her counselor no matter what orientation you work from (see ethical checklist in Chapter 3). I suggest utilizing Rogerian philosophy throughout the counseling process. Here are some further techniques:

1. *Setting the counseling environment.* The purpose of setting the environment is to allow the counselor and client to get to know one another. The counselor gathers information from the client and creates an atmosphere in which the client feels safe, comfortable, and confident about the counselor's abilities (Egan, 1994). For example:

   • Give the client choices (e.g., where to sit).

   • Mention and discuss topics of interest to client.

   • Have no physical barriers between you and your client (e.g., no desk).

2. *Confirmation of confidentiality.* It is very important that you clarify your policy on confidentiality. Scissons (1993) points out that you must ensure the client understands this policy. At times, you may have to interject this information when the client appears ready to reveal a specific matter. For example:

   COUNSELOR: Everything you say to me is confidential, unless it involves information that is dangerous to yourself or others. Do you understand?

3. *Consent.* Consent may include discussion about length of sessions, fees, and counselor qualifications. Consent means that the counselor has the client's permission to discuss a problem and has a commitment from the client to pursue the counseling process. Scissons (1993) speaks of gaining validation of the client's willingness to change. For example:

COUNSELOR: We have completed our introductions and explained confidentiality. Do you want to continue the counseling process?

4. *Active listening.* This refers to the counselor's ability to interpret the underlying meaning of what the client is saying. Egan (1994) describes the importance of reading the client's nonverbal as well as verbal communication. This includes taking into account the client's body language (gestures, posture), facial expressions (smiling, frowning), voice-related behavior (tone, level, emphasis), physiological responses (pupil dilation, blushing), physical characteristics (weight, fitness) and general appearance (grooming). All of these factors enter into the process of actively listening to a client and accurately hearing what the client is saying, trying to say, or not saying.

   Cavanagh (1990) emphasizes the active portion of the listening process. A common complaint heard about counselors is: *He just sat there and listened, nodded a few times, repeated what I said, and told me when my time was up* (p. 86). Good listening involves interaction; it requires participation on the part of the counselor. It goes hand in hand with stimulating and encouraging the client to respond spontaneously to the counselor (Cavanagh, 1990). For example:

CLIENT: He never does what he says he will!

COUNSELOR: You find it difficult to get a commitment from him?

CLIENT: Yes, he promises he will change but never does.

COUNSELOR: (nods) It sounds like he has made many empty promises to change.

5. *Reflection.* One of the fundamental counseling skills needed by a professional is clarification of the client's remarks. Clarifying helps avoid ambiguity and vagueness. There should be no assumption on the part of the counselor of automatically understanding the client's concerns, or that the client understands the counselor's intent. Rogers (1965a) referred to this process as *reflection,* whereby the counselor reflects statements back to the client in order to ensure that the client has been understood, or in order to give the client insight into what has been expressed to the counselor. For example:

   • What I hear you saying is . . .

   • You are very distraught. Is that the feeling behind what you have said?

   • Are you telling me that you think . . .

Summary

Carl Rogers was a caring, sensitive individual whose personality exuded the therapeutic process he so successfully propagated. I believe there are only a few people who can do Rogerian therapy as well as Rogers himself. Beneath his personality lay a sincere desire to be as consistent a humanist as possible. He built his therapy system upon a belief in an innate self-actualizing tendency in every human being. Today, his core conditions have been adopted by a wide variety of theorists and counselors, whether humanist or otherwise.

The professional counselor will find it very beneficial to work these principles into a personal counseling style, particularly in order to set a good counseling environment and to build a therapeutic relationship with the client. Though many people, both counselors and clients, have indicated a need for more direction than Rogers was willing to provide, such directiveness is normally only effective once trust has been established. The person-centered approach is ideal for building that trust.

References

Cavanagh, M. E. (1990). *The counseling experience.* Prospect Heights, CA: Waveland Press.

Corsini, R. J. (Ed.) (1994). *Encyclopedia of psychology* (2nd ed.). New York: Wiley.

Corey, G. (1996). *Theory and practice of counseling and psychotherapy* (5th ed.). Pacific Grove, CA: Brooks/Cole.

Egan, E. (1994). *The skilled helper: A problem management approach to helping.* Pacific Grove, CA: Brooks/Cole.

Gazda, G. M. et al. (1991). *Human relations development: A manual for educators.* Boston: Allyn & Bacon.

Goodyear, R. K. (1987). In memory of Carl Ransom Rogers (editorial). *Journal of Counseling and Development* 65:523–24.

Hampton-Turner, C. (1981). *Maps of the mind.* New York: Collier.

Hunt, M. (1993). *The story of psychology.* New York: Doubleday.

Nordby, V., Hall, C. (1974). *A guide to psychologists and their concepts.* San Francisco: W. H. Freeman.

Rogers, C. R. (1961). *On becoming a person: A therapists view of psychotherapy.* Boston: Houghton Mifflin.

Rogers, C. R. (1965a). *Client-centered therapy.* Boston: Houghton Mifflin.

Rogers, C. R. (1965b). Part one. In E. Shostrom (ed.), *Three approaches to psychotherapy* (film). Santa Ana, CA: Psychological Films.

Rogers, C. R. (1970). *Carl Rogers on encounter groups.* New York: Harper & Row.

Rogers, C. R. (1986). Client-centered therapy. In I. L. Kutash, A. Wolf (eds.), *Psychotherapist's casebook.* San Francisco: Jossey-Bass.

Rogers, C. R., Raskin, N. J. (1989). Person-centered therapy. In R. J. Corsini, D. Wedding (eds.), *Current psychotherapies* (4th ed.). Itasca, IL: F. E. Peacock.

Rogers, C. R., Sandford, R. C. (1985). Client-centered psychotherapy. In H. I. Kaplan et al. (eds.), *Comprehensive textbook of psychiatry* (4th ed.). Baltimore: William & Wilkins.

Saulnier, M. (1971). *Counseling.* Ottawa: Crown copyright.

Scissons, E. H. (1993). *Counseling for results: Principles and practices of helping.* Pacific Grove, CA: Brooks/Cole.

Thorne, B. (1992). *Carl Rogers.* Newbury Park, CA: Sage.

Todd J., Bohart, A. C. (1994). *Foundations of counseling and clinical psychology.* New York: Harper-Collins.

Chapter 9

More Directive   Less Directive

Jungian Philosophy: Carl Jung

Jung's patients were not sexually inhibited but religiously troubled, and his theories focused on philosophical, cosmic, and religious themes.

Maxmen, 1995

HISTORY

Carl Jung (1875–1961), a Swiss psychoanalyst, was a major contributor to the field of analytical psychotherapy. He was both an innovative, creative practitioner and a bold thinker. The son of a minister, during his formative years Jung experienced a great deal of confusion as to modern German theology and religion. This struggle led to his interest in psychology. In 1900, Jung received his medical degree from Basel University. During these early years, Jung was influenced by Freud, who appointed him the first president of the International Psychoanalytic Association. Their parting of ways took place after the publication of Jung's book *Symbols of Transformation* (1911), which interpreted the libido as general psychic energy rather than as sexual energy, thereby significantly dethroning sexuality as the all-encompassing causative element in things psychic.

In addition to being a medical doctor and psychiatrist, Jung was also a painter, cook, sailor, stonemason, and dreamer. Throughout his life he studied religion, mythology, philosophy, and mysticism, in a dozen languages. Both counseling and the role of the professional counselor are enhanced by Jung's spiritual philosophy, with its reference to a higher power and its belief in the importance of meaningfulness as a quest for humankind.

Jung practiced analytical psychotherapy and wrote extensively about analysis. This chapter presents a very brief description of his philosophy. It is important for the professional to recognize that exploration and work with the unconscious is the driving force of analytical psychotherapy (Kaufmann, 1989) and it requires extensive training.

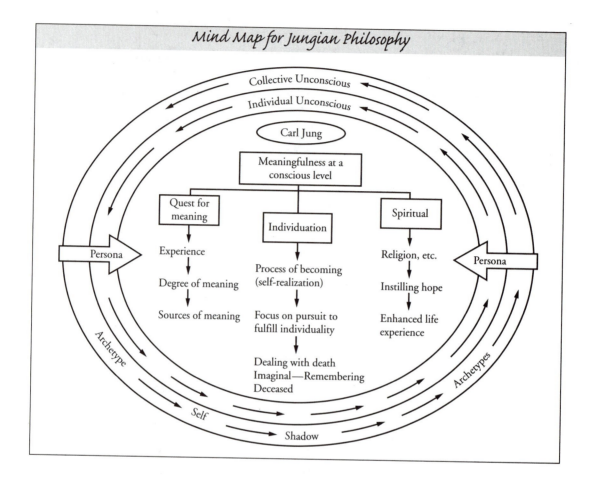

MAIN IDEAS

I have included Jung's work to provide you with an opportunity to explore spirituality within the counseling process. Carl Jung incorporated a definitive spiritual component into his philosophy of psychoanalysis. His orientation makes an attempt to address the spiritual dimension of counseling and may help the professional who is interested in taking into account mind, body, and *spirit*. Jung's spiritual component enhances the concept of *holism* in counseling.

In Jung's philosophy, the unconscious and conscious are two complementary components of the psyche. Jung defined the ideally healthy human personality as one that has achieved a balance between the conscious and the unconscious, between interior and exterior life. The unconscious and conscious also have a universal dimension in the philosophy of Jung. In the following discussion, visualize yourself traveling inward from the universal to the inner psyche of the individual.

Universality

Universal aspects of Jungian philosophy include the following:

- *The collective unconscious.* Visualize the collective unconscious as the outer circle. The collective unconscious is that part of the unconscious which is universal. Kaufmann describes it as the "larger totality."

- *The unconscious.* As we move from the collective unconscious inward, we enter the unconscious, which is a "psychic sphere" that contains the individuals energy.

- *Non-personal unconscious (and personal unconscious).* Jung divided the unconscious into two components. Again, visualize movement from universality inward. The first level is the non-personal conscious, which contains the past, present, and future psyche of mankind, along with inherited predispositions. The second level is the personal unconscious, representing the individual's life and everything experienced in that individual's lifetime.

- *Archetypes.* The collective unconscious is connected to the individual psyche through archetypes. An archetype is a definite potential within the individual; it is activated by the experience of that person's life. We may visualize archetypes as universal designs we receive from the collective unconscious; these archetypes are activated by our individual experience. Jung named some of these archetypes. For example, Kaufmann (1989) explains that the *persona* is the archetype of *adaptation,* which mediates the inner psyche and outside world (see mind map). The *shadow* archetype Jung described as that part of ourselves we do not like. We may see a characteristic in others that we dislike; but, without realizing it, we also have this trait. Another archetype Jung named was that of *self,* which he believed to be the individual's expression of meaning in life. This may be illustrated by the artist whose true self is realized through pursuit of and participation in the creative process (e.g., painting a picture).

  We have now traveled from the collective unconscious to the individual unconscious.

- *Personality types.* In Western culture, Jung has been lauded as the originator in psychology of the personality types *extrovert* and *introvert.* The extrovert focuses primarily outside the self and upon objective events. The extrovert may be outgoing and sociable. By contrast, the introvert focuses inward and tends to see things from a personal perspective (i.e., subjectively); this individual feels comfortable spending time alone and may choose solitary pastimes. These personality types occur in combination (Jung, 1961), but this basic differentiation has been used extensively throughout the field of psychology. One example is the Myers-Briggs Type Indicator, a psychological instrument based on Jung's personality types that is designed to measure personality dimensions (e.g., extroversion and introversion) (Briggs-Meyers, 1987).

Individuality

Jung, having identified what is universal to human beings, turned to a consideration of individuality.

- *Individuation.* In terms of human motivation, or that which induces us to do what we do, Jung theorized that along with sex, aggression, hunger, and thirst, *individuation* is an instinct. This term he described as an inner force that drives us to achieve individual wholeness in our lives.

- *Mid-life crisis.* Storr (1988) states that, in the actual practice of psychotherapy, the major contribution of Jung's work concerns adult development. Most of Jung's patients were middle-aged, and Storr relates that Jung was the first psychiatrist to draw attention to what is now familiarly known as the mid-life crisis. In the 1990s, the correlation of middle age with a search for meaning in life is not an unfamiliar scenario.

- *Death.* Jung also dealt extensively with the subject of death in both his philosophy and personal life. He believed the dead remain with us through the images we have of them and that this image-filled process is both positive and healing. Jung believed in the immortal soul and viewed death as the ending of the physical state in which the psyche has lived (Segaller, 1989).

- *Spirituality.* Carl Jung always referred to the God-image to illustrate his belief that the interpretation of God is very individualistic. He believed the spiritual component to be inherent in the individual and stated that without religion, mythology, or spiritual dimension the individual was incomplete, without meaning. For the field of psychotherapy, Jung legitimized spiritual experience, giving it leadership in the search for meaning. The human services counselor who knows of the dimensions that Jung incorporated into his philosophy (e.g., religion, quest for meaning) may recognize these in their clients.

- *Relationships.* Jung was open in his interpretation of sexuality. He considered both men and women to be potentially bisexual. In his philosophy, a female has an unconscious *animus* (masculine) aspect and a male has an unconscious *anima* (female) aspect.

The Dream

As a practitioner, Jung carried out intensive dream interpretation with his clients. His work with dreams was the core of his psychoanalysis. In his study of dreams, Jung concluded that the unconscious is autonomous; therefore, we do not choose the content of our dreams. The symbols of our dreams, he believed, may be mysterious, but hold the key to regaining balance in our lives. Segaller (1989) describes Jung's belief that dreams are a source of wisdom that can be transformed into positive benefit to the individual.

Dream interpretation is beyond the scope of this book. However, the human services counselor may appropriately refer a client to the writings of Carl Jung for bibliotherapy (e.g., the counselor assigns the client homework that involves appropriate readings). This should not be done unless the counselor has become acquainted with Jung's work. One text that may be of interest is *Memories, Dreams, Reflections,* written by Jung in 1961 (see References).

TECHNIQUES

Carl Jung's philosophy is expansive and challenging. His concept of the collective unconscious integrates a spiritual dimension with human psyche. The Jungian concepts of *quest for meaning, individuation,* and *spirituality* allow for depth, holism, and direction in the counseling process.

We have formulated three categories, located at the center of the mind map, to assist you in incorporating a spiritual component into your counseling orientation. The persona archetype has been included in the mind map as the facilitator, the doer, enabling the client to carry out change. The following are examples of how to integrate these categories into a counseling paradigm.

1. *Quest for meaning.* The counselor may approach this idea with clients through discussion of their experiences, the extent to which the experiences are meaningful for them, and the sources of meaning in their life. This may lead to a discussion of the types of experiences that could provide more meaning in life. For example:

   COUNSELOR: In what way does this activity add meaning to your life?

   COUNSELOR: Can you think of other sources of information that could assist you in your quest for meaning?

   COUNSELOR: If you can't see yourself doing anything meaningful, can you think of anyone you know who seems to really enjoy something in life? What do they enjoy doing? What meaning might it provide for them?

2. *Process of individuation.* The counselor may discuss with the client what aspects of individuality is desired. These aspects may include career, leisure, educational, or philosophical pursuits and personal growth. In the area of personal growth, the client may wish to discuss the subject of death. This may include exploration of the individual's fear of dying or loss of a loved one. The following cluster of questions is applicable to each of the categories of career, leisure, education, and philosophical growth. For example:

   COUNSELOR: Is there a particular activity, belief, or pastime you have always wanted to pursue?

   COUNSELOR: In what way is this career going to help you express your individuality?

   COUNSELOR: Do you know a person in this career you want to be like? What are the traits of this person? What does this person do? What is it about this person's life that is so appealing to you? In what way will this contribute to your individuality?

   Here are some examples pertaining to personal growth:

   COUNSELOR: Can you think of some way in which you would like to grow as a person?

COUNSELOR: Is there anything you have always wanted to do? How would this contribute to your personal growth?

COUNSELOR: I can see that this subject is one about which you have been thinking a great deal. Since you read a lot, I can suggest some readings that might help you to work out a personal philosophy on this subject.

Authors with whom you, as a professional counselor, may wish to become acquainted are: Rollo May, *Freedom and Destiny* (1981); Elisabeth Kübler-Ross, *Death: The Final Stage of Growth* (1975); and Carl Jung, *Memories, Dreams, Reflections* (1961).

3. *Spirituality.* The human services counselor may introduce spirituality as a suggestion, as an option, or as a vehicle of hope and courage. Bibliotherapy—for example reading the Bible or scriptures of other world religions—may be suggested in Jung's philosophy. The following are examples of questions that could be used to facilitate spiritual exploration:

COUNSELOR: Some people find spiritual activity (like going to church) helpful. Might that be a possibility for you?

COUNSELOR: Sometimes spirituality can be helpful in that it gives hope and courage. Have you ever considered exploring that direction?

COUNSELOR: You appear to have an interest in spirituality; can you think of particular aspects you might be interested in, aspects that may be helpful to you?

Summary

The work of Jung may be helpful to the professional counselor when addressing the concepts of individuation, a quest for meaning, or spirituality.

It is difficult for us to fully appreciate what a bold move it was for Jung to orient his psychoanalysis toward the spiritual at a time when the leading intellectual climate of his culture was so decidedly uninterested in it. His great contemporary, Adler, whose psychology has also been described as having a religious tone (Adler, 1958; Mosak, 1987) and whose social interest theory also stood in a human-oriented religious tradition (Mosak, 1989), was nonetheless definitely secular in his adaptation of that tradition. Jung; however, went further, claiming both Adler and Freud were incomplete (Jung, 1956) and presented his thoughts about the unconscious based on research in a wide range of mystical and religious literature.

References

Adler, A. (1958). *What life should mean to you.* New York: Capricorn.

Briggs-Myers, I. (1987). *Introduction to type.* Palo Alto: Consulting Psychologist Press.

Jung, C. G. (1956). *Two essays on analytical psychology.* New York: Meridian.

Jung, C. G. (1961). *Memories, dreams, reflections.* New York: Vintage.

Jung, C. G. (1964). *Man and his symbols.* Garden City, NY: Doubleday.

Kaufmann, Y. (1989). Analytical psychotherapy. In Corsini, R. J., Wedding, D. (eds.), *Current psychotherapies,* 4th ed. Itasca, IL: F. E. Peacock.

Maxmen, J. S. (1995). *Essential psychopathology and its treatment.* New York: Norton.

Middelkoop, P. (1989). *The wise old man (healing through inner images).* Boston: Shambhala.

Mosak, H. H. (1987). Religious allusions in psychotherapy. *Individual Psychology 43:*496–501.

Mosak, H. H. (1989). Adlerian psychotherapy. In R. J. Corsini, D. Wedding (eds.), *Current psychotherapies,* 4th ed. Itasca, IL: F.E. Peacock.

Nordby, V., Hall, C. (1974). *A guide to psychologists and their concepts.* San Francisco: Freeman.

Reader's Digest (1990). *ABCs of the human mind: A family answer book.* New York: Reader's Digest Association.

Segaller, S. (1989). *The wisdom of the dream.* Boston: Shambhala.

Storr, A. (1988). *Solitude: A return to self.* New York: Free Press.

Chapter 10

More Directive   Less Directive

Existential Therapy: An Introduction

Choice alone appears as a choice between objectivities; freedom is the choice of my own self.

Jaspers, 1970

HISTORY

Existential therapy has a long and varied history, its roots dating back to the religious writer Sören Kierkegaard in the 1840s and the philosopher-poet Friedrich Nietzsche in the 1870s and 1880s. These writers, plus Edmund Husserl, who developed the method of phenomenology in the first two decades of this century, had a profound influence on German philosophers in the 1920s and 1930s. They, in turn, influenced certain psychologists to adopt existential philosophical concepts as an alternative to traditional psychoanalysis (May, 1958).

In the 1950s, these German and French works were translated into English, finding fertile ground in the minds of humanistic psychologists Maslow, Rogers, and Allport (May et al., 1960). In the last forty years, some original work has been done by May, Frankl, and Yalom. Key players in the development of existential therapy and phenomenology and presented in Figure 10–1.

Existential therapy had developed extensively throughout this century. It holds appeal for those whose orientation is humanistic, person-centered, and meaning-focused. This text will look at the last three modernday existential psychologists: Frankl, May, and Yalom.

MAIN IDEAS

Existential therapy has its underpinnings in existential philosophy. The dawning of a new millennium tends to bring existential psychology to the fore as you develop your counseling orientation because, as the century ends, many people are becoming more cognizant of their own mortality. They are searching for the meaning of life, personal freedom, choice, responsibility, anxiety, and guilt. This chapter provides a paradigm for addressing some of these issues with your clients and yourself.

- S. Kierkegaard (1813–1855) A religious, philosophical, and psychological writer; he rejected determinism and identified the major concern among humans as despair and spiritlessness. Kierkegaard believed the road to an authentic self is a journey of individuation and differentiation by a person from his/her social culture (Corsini, 1994). Most of the now-standard existential themes (anxiety, freedom, choice, subjectivity, etc.) are derived from Kierkegaard (May, 1958).

- F. Nietzsche (1844–1900) A philosopher and poet; he taught that human motives are found in our instincts and drives, not thought and reason; consciousness and conscious acts are done in service to untamed drives (Corsini, 1994). The creative will holds a central place in his deeply psychological philosophy (Nietzsche, 1954).

- E. Husserl (1859–1938) A phenomenologist, he was the founder of the modern phenomenological method, which emphasizes a descriptive (rather than conceptual) approach to the contents of consciousness, and the intentionality of all conscious acts (Husserl, 1962). It focuses on the individual's subjective interpretation of situations rather than on an objective, cause-and-effect approach.

- M. Scheler (1874–1928) A philosopher; Scheler was one of the original innovators of phenomenology. He provided profound insights into the psychology of ethical valuing, and of feeling (Spiegelberg, 1972).

- K. Jaspers (1883–1969) A philosopher and psychiatrist; Jaspers wrote perhaps the first book, *General Psychopathology* (1913), that introduced pathological states and an existential philosophical underpinning, the latter being developed fully in his three-volume *Philosophy* (1932). His existential explanations (elucidations) of boundary situations are considered classic in the works of existentialism (Yalom, 1980). Jaspers, combining his psychiatric training with his existential philosophy, was really concerned about how humans address problems such as death, conflict, suffering, and anxiety (Corsini, 1994).

- M. Heidegger (1889–1976) A philosopher; he wrote *Being and Time* (1927), the most influential book in the formation of existential analysis (Binswanger and Needleman, 1963; Boss, 1963). Heidegger, a student of Husserl, adapted phenomenology for existential concerns, created much of the terminology now standard in existential writings, and wrote extensively on being-in-the-world. He is thought to be the main bridge between existential philosophy and existential psychology (Corsini, 1994).

- J. P. Sartre (1905–1980) A philosopher; he wrote *Being and Nothingness* (1943), the most important work in French existentialism. The entire book consists of psychologically penetrating analyses and, in it, Sartre worked out his own notion of an existential psychoanalysis (Sartre, 1953).

(continued)

Figure 10–1.
(continued)

- L. Binswanger (1881–1966) A psychologist originally trained as a Freudian psychoanalyst, Binswanger sought a less confined and reductive philosophical orientation for psychology. He found this in the phenomenological approach and Heidegger's existential analysis. Binswanger helped integrate philosophy with psychiatry by transforming the existential foundations of Heidegger's work into applicable psychological terms (May, 1958).

- E. Minkowski (1885–1972) A psychologist considered the pioneer in phenomenological psychiatry. Minkowski's investigations into the disturbance of the pathological individual's relation to time were highly influential (May, 1958).

- M. Boss (1903–1990) A psychologist; originally influenced by Freud and Jung, Boss was introduced to Heidegger's work by Binswanger, and his work has the "active and apparently unqualified support of Heidegger" (Spiegelberg, 1972). Boss has "stressed man's freedom, denying all causality" (Corsini, 1994).

- V. Frankl (1905–1998) A psychologist originally influenced by Freud and Adler, Frankl found their sphere too constrictive regarding spiritual and ethical matters, so he developed his own existential analysis called *logotherapy* (meaning-therapy). Partly related to his heroic struggles in Auschwitz, and partly due to the originality of his existential insights and applications, Frankl's *Man's Search for Meaning* (1963) is well known and highly influential (Yalom, 1980).

- R. May (1909–1994) A psychologist; May was primarily responsible for introducing existential psychology to America with the publication of *Existence* (1958). May's *Man's Search for Himself* (1953), *Love and Will* (1969), and *The Courage to Create* (1975) are original contributions and developments of existential thematics (Yalom, 1980).

- I. Yalom (b. 1931) A psychologist, he wrote *Existential Psychotherapy* (1980), one of the most thorough and comprehensive texts on the subject written in English. Yalom's clinical experience and openness to empirical testing of hypotheses lend credence to his work. Yalom's four ultimate existential concerns are death, freedom, isolation, and meaninglessness (Yalom, 1980).

Prominent thinkers have defined and expanded upon *existential philosophy,* and the related *phenominology,* in a myriad of ways.

1. The following terms, provided by the Corey (1996) *Student's Manual,* will be helpful in understanding existential therapy.

- *Existential guilt.* The result of, or the consciousness of, evading the commitment to choosing for ourselves.

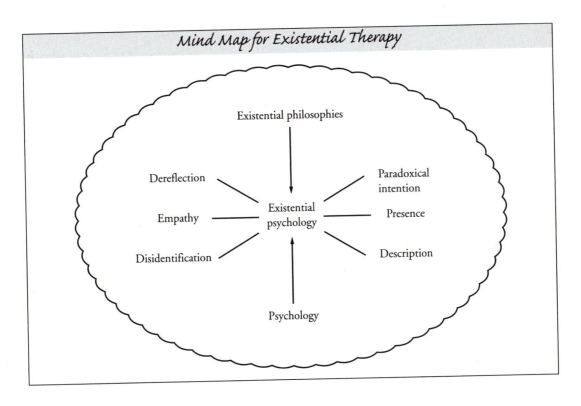

Mind Map for Existential Therapy

- *Existentialism.* A philosophical movement stressing individual responsibility for creating one's ways of thinking, feeling, and behaving.
- *Existential neurosis.* Feelings of despair and anxiety that result from inauthentic living, a failure to make choices, and an avoidance of responsibility.
- *Existential vacuum.* A condition of emptiness and hollowness that results from meaninglessness in life.
- *Freedom.* An inescapable aspect of the human condition, implying that we are the author of our life and therefore are responsible for our destiny, and accountable for our action.
- *Phenomenology.* A method of exploration that uses subjective human experiencing as its focus. The phenomenological approach is a part of the fabric of existentially oriented therapies, of Adlerian therapy, of person-centered therapy, of gestalt therapy, and of reality therapy.
- *Restricted existence.* A state of functioning with a limited degree of awareness of oneself, and being vague about the nature of ones problems.

2. Heidegger (1962) says "Phenomenology means. . . to let that which shows itself be seen from itself in the very way in which it shows itself from itself."

3. Jaspers (1963) states "Phenomenology sets out on a number of tasks: it gives a concrete description of the psychic states which patients actually experience and

presents them for observation. . . . Since we never can perceive the psychic experiences of others in any direct fashion, as with physical phenomena, we can only make some kind of representation of them. There has to be an act of empathy, of understanding."

4. May (1983) teaches that existentialism is not a comprehensive philosophy or way of life, but an endeavour to grasp reality. Existentialists are centrally concerned with rediscovering the living person amidst the dehumanization of the modern world.

5. Husserl (1962) teaches that conscious acts are intentional; that is, a certain type of conscious act (thinking, feeling, valuing) is determined by an intention (a purpose or guiding meaning) that is projected by someone, about some object(s), in some manner. This intention sorts and organizes all sensory input into recognizable perceptions that have meaning for us within a certain already established "world," or context of meanings (our own particular paradigm).

6. The kind of connection between self and world or man and nature is seen, in existentialism, as dependent on one's attitude toward death (e.g., whether death is viewed as part of the self, or externalized, as a threat from the external world) (Heidegger, 1962).

7. May (1958) says the "world is the structure of meaningful relationships in which a person exists and in the design of which he participates."

8. May (1983) states "For no matter how interesting and theoretically true is the fact that I am composed of such and such chemicals or act by such and such mechanisms or patterns, the crucial question always is that I happen to exist at this given moment in time and space, and my problem is how I am to be aware of the fact and what I shall do about it."

9. Laotzu (in May, 1983) states "Existence in infinite, not to be defined; and though it seem but a bit of wood in your hand, to carve as you please, it is not to be lightly played with and laid down."

10. May (1960) believes the possibility of death jars us loose from the treadmill of time becuse it so vividly reminds us that we do not go on endlessly.

11. For Yalom (1980), "the study of psychopathology is the study of failed death-transcendence." He asserts that death-transcendence is a major motif in human experience—from the most deeply personal internal phenomena, our defenses, our motivations, our dreams and nightmares, to the public macro-societal structures, our monuments, theologies, ideologies, our stretch into space, indeed our entire way of life, our filling time, our addiction to diversions, our unfaltering belief in the myth of progress, our drive to get ahead, our yearning for lasting fame.

12. Existentialists (Kierkegaard, 1957; Heidegger, 1949) distinguish fear from anxiety. They state that, in experiencing fear, we dread some definite thing (external), while, experiencing anxiety, we dread no thing; we stand powerless before nothing, a nothing that comes from nowhere in the external world, but from the heart of our being.

13. Yalom (1980) says we combat "anxiety by displacing it from nothing to something." As May (1977) puts it, "anxiety seeks to become fear" and thus manageable in some way. Thus, we attempt to turn our anxiety into fear by externalizing it and projecting it onto something concrete.

14. Yalom (1980) concludes that we develop two basic defense mechanisms against death-awareness, both of which lead to psychopathology or existentially maladaptive behavior. He calls these: *specialness,* or the inviolability (not to be broken) of personal security (protection from physical, emotional and psychological violations), as exemplified by workaholism, narcissism, aggression and control, and compulsive heroism; and *belief in an ultimate rescuer,* a personal god, a leader, a cause, a doctor, or a lover who will protect us.

15. The openness to the certainty of death is a *resolute anxiety,* a commitment to truth that accepts and affirms, without knowing why, the apparent internal necessity of existential self-sacrifice (Heidegger, 1949).

16. Our quest for immortality is a prime example of our denial of being-towards-death. To be authentic, we need to embrace the actuality of our mortality and begin to fully live. For we cannot truly embrace life until we fully accept its nemesis—death (May, 1958).

17. Freedom is finite, and bears its own responsibility for choice and authenticity (Heidegger, 1962).

18. In choosing to be the finite beings that we are, we come to ourselves here and now, in historic singularity (Jaspers, 1970). We exist, transparently, in the unique moment-of-vision once and for all; we have been and we have seen and now we affirm our existence (Heidegger, 1969).

19. May (1958) says "The aim of the therapy is that the patient experiences his existence as real. The purpose is that he become aware of his existence fully, which includes becoming aware of his potentialities and becoming able to act on the basis of them.

If you choose to use existential therapy in your counseling orientation, you may find it useful to address the following types of client concerns:

- Finding a purpose in life and learning who you are
- Understanding that we all create our own destiny
- Recognizing that life is challenging and the anxiety that comes with it is predictable as well as manageable. People need to learn they are not alone in their fears (e.g., addressing one's own dark side is a common challenge).
- Creating one's own purpose and life direction
- Determining the rules one will live by and setting one's life goals

These are only examples; the core themes are all in the quest for meaning and becoming genuine as a person. As the new century becomes a reality, I find that more and more clients are becoming focused on these types of concerns.

TECHNIQUES

It is important to understand that existential therapy is not really a set of techniques. Existential therapy is based on human experience, and could be characterized as an attitude of being present with your clients to help offset their anxiety as they wrestle with life issues and choose the necessary steps for a new beginning. Nevertheless, a number of terms have arisen in the practice of existential therapy:

1. *Dereflection* (Frankl, 1955). Diversion of clients' attention away from their problems and toward accessible meanings outside of the problematic self. This is a way of assisting clients to understand that they can choose to be psychologically healthy and active. For example, clients who, perhaps, have perfectly justified complaints about loneliness may be asked to avoid dwelling on them, and instead notice how creative they are in difficult situations, or how responsibly they act (e.g., choosing to avoid people at a party).

2. *Paradoxical intention* (Frankl, 1955). Intentional exaggeration of problematic behaviors, which breaks the compulsion to do them. For example, a student who will not study, and is in a power struggle with parents, is sent to a counselor. The counselor encourages the student to continue not, under any circumstances, to study. The parents' and counselors' support of the student's plan dissolves the motivating rebellious energy and the student will now have no conflict except that of not passing. The student without conflict has the opportunity to see the value of passing in school and begins to study.

3. *Disidentification* (Bugental 1973). One of the results of avoiding death is an over-identification with one's roles in the world. This exercise undoes the identification and generates powerful emotions (Yalom, 1980) as clients approache the self-discussion and confusion of their own being. For example, clients write, on separate cards, eight important answers to the question Who am I? They then arrange them in order of importance, the top card being the least important. The client takes the top card and is asked to contemplate, for a few minutes, what it would be like to give up that attribute. The the client takes the next card and repeats the procedure until all eight attributes have been divested (Yalom, 1980).

4. *Description* (Jaspers, 1963). The initial verbal expressions of the clients' experience are best understood when they are *descriptive,* not conceptual or judgmental, since description "presents" the psychic state, and explains the client's position at the moment. For example:

CLIENT: I've concluded that life is meaningless because for me there has never been good proof that God exists.

COUNSELOR: When you say *meaningless,* describe what this means for you.

CLIENT: I feel I have no purpose. What are we, or why am I here? I feel as if I am lost in the fog.

COUNSELOR: I now feel I understand what you are saying. Could you be more specific, please?

5. *Empathy* (Jaspers, 1963). Empathy is expressed when the counselor provides a degree of understanding that grows out of listening to the client's experience as if from the client's point of view. This act of empathy, like description, comes before all observation. For example, in the preceding dialogue the counselor expresses empathy both when asking the client to describe personal experience (rather than thoughts) and when saying "Now I feel I understand." For further examples of empathy, see Chapter 8.

6. *Presence* (May, 1958). Therapeutic presence means *being there,* being authentically oneself, open and truly with the client, and encountering the client as a human being. As this is not, strictly speaking, a technique, but is in fact often antithetical to techniques, we give no example here but emphasize that therapeutic presence is a function of the degree of personal authenticity of the therapist. However, we again suggest the reader review Chapter 8, in which we discussed setting the counseling environment.

Summary

This chapter provides you with an historic look at existential therapy—where it began and how it may be used in modernday counseling. We have addressed existential therapy, not because of its vast content and techniques, but because, no matter what your counseling orientation, when you clear everything away all that is left is the person who needs help and you. As a human services counselor, you cannot forget that you are a professional counselor; however, in the end, we all all just human beings trying to find meaning in this life.

References

Assagioli, R. (1976). *Psychosynthesis.* New York: Penguin.

Binswanger, L., Needleman, J. (1963). *Being-in-the-world.* New York: Basic Books.

Boss, M. (1963). *Psychoanalysis and daseinanalysis.* New York: Basic Books.

Bugental, J. (1965). *The search for authenticity.* New York: Holt, Rinehart, & Winston.

Bugental, J. (1973). Confronting the existential meaning of my death through group exercises. *Interpersonal Development 4:*1948–63.

Bugental, J. (1976). *The search for existential identity.* San Francisco: Jossey-Bass.

Corsini, R. J. (1994). *Encyclopedia of counseling, Vol. 1* (2nd. ed.). New York: Wiley.

Frankl, V. (1955). *The doctor and the soul.* New York: Knopf.

Frankl, V. (1963). *Man's search for meaning.* New York: Pocket Books.

Heidegger, M. (1949). *Existence and being.* Chicago: Gateway.

Heidegger, M. (1962). *Being and time.* New York: Harper & Row.

Heidegger, M. (1969). *The existence of reasons.* Evanston, IL: Northwestern University Press.

Husserl, E. (1962). *Ideas.* New York: Collier.

Jaspers, K. (1963). *General psychopathology.* Chicago: University of Chicago Press.

Jaspers, K. (1970). *Philosophy, Vol. 2.* Chicago: University of Chicago Press.

Kierkegaard, S. (1941). *Concluding unscientific postscript.* Princeton: Princeton University Press.

Kierkegaard, S. (1957). *The concept of dread.* Princeton: Princeton University Press.

May, R., Angel, E., Hienberger, H. (Eds.) (1958). *Existence: A new dimension in psychiatry and psychology.* New York: Basic Books.

May, R. (1960). *Existential psychology.* New York: Random House.

May, R. (1969). *Love and will.* New York: Dell.

May, R. (1975). *The courage to create.* New York: Bantam.

May, R. (1977). *The meaning of anxiety.* New York: W.W. Norton.

May, R. (1983). *The discovery of being.* New York: W.W. Norton.

Nietzsche, F. (1954). Thus spake Zarathustra. In Kauffman, W. (ed.)., *The portable Nietzsche.* New York: Viking.

Perls, F. (1969). *Gestalt therapy verbatim.* Lafayette, CA: Real People Press.

Sartre, J. P. (1953). *Being and nothingness.* New York: Philosophical Library.

Spiegelberg, H. (1972). *Phenomenology in psychology and psychiatry.* Evanston, IL: Northwestern University Press.

van Deurzen-Smith, E. (1988). *Existential counselling in practice.* London: Sage.

Yalom. I. (1980). *Existential psychotherapy.* New York: Basic Books.

Chapter 11

Adlerian Psychotherapy: Alfred Adler

Humans are constantly "becoming"—moving toward fictional goals that they think lead to superiority. At times such behavior is self-defeating.

Adler, 1956

HISTORY

Alfred Adler (1870–1937) was born in Vienna and died 67 years later while on a lecture tour in Scotland. Adler obtained his medical degree in 1895 from the University of Vienna. He practiced ophthalmology before becoming a psychiatrist. He was a charter member of the Vienna Psychoanalytic society until he severed ties with Freud in 1911. After serving as a medical doctor in the Austrian army during World War I, Adler became interested in child guidance. In 1935, he settled in the United States, working both as a psychologist and psychiatrist (Nordby and Hall, 1974). After Adler's death in 1937, Rudolf Dreikurs was the most significant figure in bringing Adlerian psychology to the United States, especially as its principles applied to education, individual and group therapy, and family counseling.

Adler developed a personality theory called individual psychology, which views the person holistically as creative, responsible, and becoming (Mosak, 1989). Adler himself wrote the following introspection:

> As I come to a fuller appreciation of my own worth, I grow in confidence, in my sense of adequacy, and in my capability to live responsibly and effectively. This personal growth contributes to an emotional and spiritual warmth which becomes part of my sense of myself and my relationship with others. (Reecel, 1993)

Roazen (1975) explains that Adler, through childhood illness, developed a feeling of inferiority, about which he would later have much to say. Adler also learned to compensate for his perceived weakness by focusing on seeking his own security and supremacy (Storr, 1988), thus laying the foundation for Adlerian psychotherapy.

For more information about Adlerian psychotherapy, write to: The North American Society of Adlerian Psychology, 65 East Wacker Place, Suite 400, Chicago, Illinois 60601.

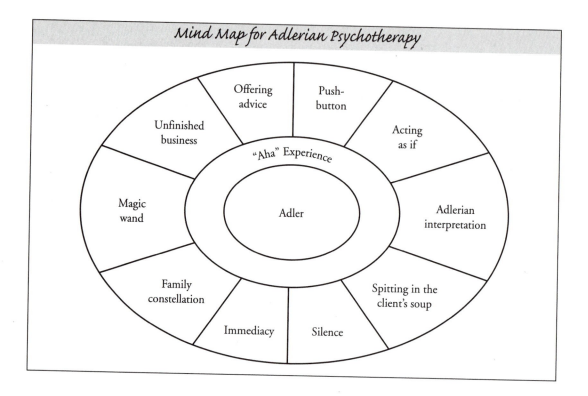

Mind Map for Adlerian Psychotherapy

Offering advice · Push-button · Unfinished business · Acting as if · "Aha" Experience · Magic wand · Adler · Adlerian interpretation · Family constellation · Spitting in the client's soup · Immediacy · Silence

MAIN IDEAS

We have included Adler's work in this text because Adlerian psychology has grass-roots concepts accessible to the human services counselor. For this reason, the following points are especially relevant to this text:

- Adler (1969) views individuals as having a need to be connected and united with others so they can learn to master the five main tasks that address life problems: building friendships, establishing intimacy, contributing to society, getting along with oneself (self-task), and spiritual tasks.

- Bernstein (1991) describes Adler's focus as oriented to the here and now. Adler believed all human beings have potential.

- Mosak (1989) states that Adlerian psychotherapy holds that children who have not achieved the recognition they think they deserve develop into individuals who are discouraged, rather than sick. Adler believed the therapeutic task was to encourage people to activate their social interest and to develop a new lifestyle through relationship, analysis, and actions.

- Adler (1963) believed personal happiness could never be obtained at the expense of others. He taught we all should consider the happiness of community and self; this allows society as a whole to benefit, as well as the individual.

- Todd and Bohart (1994) indicate Adler was perhaps best known in the field of psychology for his concepts of inferiority complex and the striving for superiority.
- Dinkmeyer, Dinkmeyer, and Sperry (1987) relate that Adler believed people were motivated primarily by social forces and were driven to obtain specific individual life goals.
- Adler (1964) asserted inferiority was not a negative factor; on the contrary, Adler believed inferiority was the motivation to master our environment.
- Dreikurs (1952) states Adler believed basic mistakes in one's lifestyle can be understood as mistakes in commonsense thinking.
- Hunt (1993) explains that Adler believed birth order within the family was a factor greatly affecting child development and, more important, the child's family position. Adler also observed the manner in which the child was reared.
- Corsini (1994) describes Adlerian therapy techniques as very flexible, allowing the counselor to incorporate a variety of other techniques, or to utilize intuitive imagination in the creation of a therapeutic approach, with the end goal of helping the client grow and move on with life in a positive manner.
- Sweeney (1989) pinpoints the need to understand from the onset the social context of the client; this is a necessary and important part of an individual's personal development and ability to find solutions.
- Adler's *fictional finalism* can be explained as an imagined goal that motivates action in our effort to obtain perfection. It is the central goal, giving direction to human behavior, and joining it to the personality (Adler, 1963).
- As long as new learning continues and the client is experiencing the desired change, Adler recommends the counselor continue whatever he or she is doing (Mosak, 1989).
- Adler believed client resistance is due to the fact the client and counselor have not yet found common goals (Adler, 1958).

TECHNIQUES

Adler (1958) explains that the overall goal of Adlerian psychotherapy is for the client to develop insight. From these insights, counselors help clients to obtain the skills needed to be more balanced and happy in life. When clients express an "Aha!" they have experienced an insight. The "Aha!" experience enables the client to gain self-confidence in confronting present difficulties. Here are some of the elements of Adlerian technique.

1. *Exploration of the family constellation.* Discussion of the family constellation enables the counselor to explore the client's family position and discover basic assumptions that are inherent in the lifestyle of the client. Mosak (1989) describes the premise of Adlerian psychotherapy as the conviction that the family creates the primary social environment. Dreikurs (1969) explains that Adler believed it

was important for the counselor to develop an understanding of the various family personalities that may be part of the client's present concerns. This helps the counselor understand what Adler described as *private logic.* Corsini (1994) defines private logic as the way in which the client "consciously experiences ideas, emotions, interests, urges, and impulses which are explained by various rationalizations." Adler believed every child searches for significance and competes for a position within the family constellation. Mosak (1989) provided the following example:

- In a family of three children, the oldest child, being the firstborn, receives attention simply through holding the primary position in the family constellation. The youngest child, being the baby and the newest member, also receives attention. The middle child, however, may receive less attention, and perceive self as less important. Adler believed this might explain why children try to assume their perceived rightful position within the family structure.

2. If clients do not reach this position they become discouraged. This may be the reason they rebel. Adlerian psychology holds that childhood is where incomplete cognitions and interpretations occur. Unless these misguided interpretations are changed, they will affect a person throughout that individual's entire lifetime. Dinkmeyer and colleagues (1987) stress the importance of client exploration of where that individual fits into the family structure. The counselor may discuss with clients how misinterpretations and lack of information have affected their lifestyle, and continue to affect presentday situations. Sample questions might include:

COUNSELOR: How many siblings do you have?

COUNSELOR: What was your family position?

COUNSELOR: What type of relationship do you have with your brothers and sisters?

3. *Acting as if.* This technique allows clients to explore alternatives to their present lifestyle. Adler (1963) states that when clients say, "If only I could. . ." (something that is not uncommon), the counselor respond by asking them to act "as if" they could. In doing this action exercise, the counselor may suggest the client try something new (e.g., smile when you say hello to people). Acting "as if" is reflected in the maxim "Fake it 'til you make it." For example:

COUNSELOR: Johnny, how about smiling and nodding when your brother speaks to you? And act as if you are really listening.

4. *Unfinished business using role play.* According to Adlerian psychotherapy, unfinished business is the result of an unresolved issue. A counselor who perceives the client to be dwelling in the past may use role play as a vehicle to explore unfinished business during the counseling process. If the client has a particular unresolved issue with a certain person, the client can use role play to work out the concern (Schultz and Schultz, 1994). In Adlerian counseling, using unfinished business as a technique enables the client to live in the present instead of the past. For example:

- Because of earlier conflict as a child, a client did not tell her mother, before she died, what a special person she was. The counselor sets up a role play using the chair technique (see Chapter 12) to assist the client in dealing with the unfinished business of telling her mother how special she was.

5. *Push button.* The benefit of this technique is to teach clients that they have control over their emotions (Sweeney, 1989). In Adlerian psychology, this method is effective with people who feel they are the victims of their own emotions (e.g., a client feels inferior and thus unable to express feelings). Clients are requested to close their eyes and think of something hurtful, producing a negative emotion. Once clients have experienced this emotion, the counselor is able to teach that they actually have control over their emotions through their thoughts, and thus they do not have to fall victim to their own emotions. For example:

COUNSELOR: Think of the day you got your exam grades (use an example of a painful emotion).

CLIENT: I failed my science exam. That makes me think how stupid I really am. I was really hurt when the teacher called me stupid in front of the class.

COUNSELOR: Can you think of something pleasurable and positive?

CLIENT: I remember when my friend and I spent a day on the beach.

COUNSELOR: When you think of this wonderful experience, how do you feel?

Note: The purpose of this technique is to show the client how thinking and feeling are tied together. When we feel bad, we can change our feelings by changing our thinking.

6. *Adlerian interpretations.* This is used by the counselor in an attempt to gain a better understanding of the client's world. The interpretations are posed in the form of open-ended questions. For example:

COUNSELOR: With your permission, I have a hunch I would like to share with you.

CLIENT: Go ahead.

COUNSELOR: From where I sit, it appears to me that . . . does that make sense?

7. *Offering advice.* This is a way of presenting alternatives and options to the client. Mosak (1989) explains that it does not matter who comes up with the suggestion, as long as clients believe in the suggestion and take ownership of it. This gives clients an opportunity to explore the many avenues available that will enable them to grow and take responsibility for their own lives. If your client is having difficulty finding a solution to a problem, you may offer a host of suggestions (join a club or group, exercise, read books, talk to people). For example:

COUNSELOR: From what I hear you saying, joining a support group could be very beneficial for you. What do you think?

8. *Magic wand.* This involves a visual exploration that allows clients to wish for anything they presently do not have, with the goal of promoting growth (Dreikurs, 1952). Clients are asked to pretend they have a magic wand in their

possession. To facilitate change, they are asked to alter anything in their life that they perceive would make a difference. Once clients come up with a wish, the counselor helps explore and clarify what this wish would look like, allowing them to obtain a clearer understanding of what they want.

COUNSELOR: If you could have one wish, what would it be?

CLIENT: To be really content.

COUNSELOR: What does contentment look like to you?

CLIENT: Oh, a family that is close, doesn't yell, lives in a nice place with good neighbors.

9. *Immediacy.* This technique is used when the counselor and the client deal with a situation within the counseling session. The purpose is to show clients that what they are doing in the present moment in therapy is similar to what they do in daily life that may be causing difficulty. This is done with the intention of showing clients that they have control over their own behavior. For example:

COUNSELOR: You appear to be frustrated, and you have just told me how dumb you think you are. When you get frustrated at home with your father, do you say the same thing to him?

10. *Silence.* Manaster and Corsini (1982) teach that one of Adler's more effective interventions is for the counselor to listen and say nothing. This allows the client time to process and integrate what is presently happening in the session.

11. *Spitting in the client's soup.* Dinkmeyer and colleagues (1987) explain that once counselors are able to determine the motivation and reward behind ineffective behaviors, they can attempt to challenge the client's thinking. This shows the client the real price being paid for the ineffective behavior. The goal is to spoil the payoff, so clients can no longer deceive themselves. For example:

CLIENT: I don't understand why my wife is on my back about coming home late. My job stresses me out and I deserve a couple of drinks after work."

COUNSELOR: Well, the drink may help with the job stress, but isn't it the stress at home that brought you here? So, how are the drinks helping at home?"

12. *Termination.* Adler believed that the entire counseling process should be used in each and every session. Counselors serve their clients most effectively by setting clear limits and a direction for the sessions. Before each session the counselor defines its expected outcome. It is unwise to introduce new topics or issues that cannot be properly addressed in the time remaining. The same rule applies in the last counseling session. The counselor should review all of the client's gains and the future actions needed to keep the new behavior in place. If the client does introduce a new issue, it may indicate a need to extend the number of counseling sessions.

Belkin (1988) proposes that, in termination of the session, the counselor needs to be mindful of the appropriate types of statements or questions with which to end, so not to confuse or bring unnecessary anxiety to the client. For example:

COUNSELOR: "Well, I think we have a number of things to discuss in our next session."

COUNSELOR: "So, our next appointment is Thursday at noon. Perhaps then we can better understand why this problem between you and your boss persists."

The ultimate goal of termination is to assist the client to gain closure on what has been accomplished to this point. It is not a permanent event; it is part of the process of human development. Endings open the door to new beginnings.

The point to remember in termination is that counseling needs to have a clear direction, with the ultimate outcome that clients be able to integrate comfortably back into society and be able on their own to live every day happily and with a sense of fulfillment.

Summary

The wealth of fresh concepts Adler brought to the field of psychology reflects his belief that humans are free, creative beings who are able responsibly to mold their own goals and move toward them. If problems occur, it is often because the person has developed feelings of inferiority. The counselor's job is not to "cure" (because the client is not sick), but to encourage, by helping the client make a change in lifestyle. Worthy goals focus around successful human relationships ("social interest").

Human services counselors will find in Adler's thoughts a supermarket of concepts upon which many other theorists have expanded. Compared to his well-known contemporaries, Freud and Jung, Adler's language is down-to-earth and his ideas have the ring of commonsense. A psychiatrist once upbraided Adler for his too-straightforward notions. "You're only talking common sense," to which Adler replied, "I wish more psychiatrists did" (Mosak, 1989).

I am particularly fond of Adler because his concept of the client's subjective reality underscores the importance of seeing the problem from the client's perspective, and cautions counselors to avoid trying to replace the client's model of the world with their own.

References

Adler, A. (1958). *What life should mean to you.* New York: Capricorn.

Adler, A. (1963). *The practice and theory of individual psychology.* Paterson, NJ: Littlefield, Adams.

Adler, A. (1964). *Social interest: A challenge to mankind.* New York: Browning/Mazel.

Adler, A. (1969). *The science of living.* New York: Doubleday Anchor.

Adler, A. (1972). *Neurotic constitution.* Freeport, NY: Books for Libraries Press.

Belkin, G. (1988*). Introduction to counseling* (3rd ed.). Dubuque: Brown.

Bernstein, D. A. (1991). *Psychology.* Boston: Houghton Mifflin.

Corey, G. (1995). *Theory and practice of group counseling* (4th ed.). Pacific Grove, CA: Brooks/Cole.

Corey, G. (1996). *Theory and practice of counseling and psychotherapy* (5th ed.). Pacific Grove, CA: Brooks/Cole.

Corsini, R. (1994). *Encyclopedia of psychology, Vol. 1*(2nd ed.). New York: Wiley.

Cumming, J. (1972). *Encyclopedia of psychology, Vol. 1.* New York: Herder and Herder.

Dinkmeyer, D. C., Dinkmeyer, D. C. Jr., Sperry, L.(1987). *Adlerian counseling and psychotherapy* (2nd ed.). Columbus: Merrill.

Dreikurs, R. (1952). *Fundamentals of Adlerian psychology.* Chicago: Alfred Adler Institute.

Dreikurs, R. (1969). Social interest: The basis of normalcy. *The Counseling Psychologist 1:* 45–48.

Hunt, M.(1993). *The story of psychology.* New York: Doubleday.

Manaster, F. J., Corsini, R. J. (1982). *Individual psychology.* Itasca, IL: F. E. Peacock.

Mosak, H. A. (1989). Adlerian psychotherapy. In R. J. Corsini, D. Wedding (eds.), *Current psychotherapies.* Itasca, IL: F. E. Peacock.

Nordby, V., Hall, C. (1974). *A guide to psychologists and their concepts.* San Francisco: Freeman.

Reecel, B. L. (1993). *Effective human relations in organizations.* Boston: Houghton Mifflin.

Roazen, P. (1975). *Freud and his followers.* New York: Knopf.

Storr, A. (1998). *Solitude: A return to self.* New York: Free Press.

Schultz, B., Schultz, S. E. (1994). *Theories of personality* (5th ed.). Pacific Grove, CA: Brooks/Cole.

Sweeney, T. J. (1989). *Adlerian counseling: A practical approach for a new decade* (3rd ed.). Muncie: Accelerated Development.

Todd, J., Bohart, C. B. (1994). *Foundations of clinical and counseling psychology.* New York: HarperCollins.

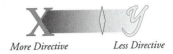

Chapter 12

More Directive   Less Directive

Transactional Analysis: Eric Berne

Transactional analysis is a humanistic, existential, dynamic theory of personality and a therapeutic modality whose practitioners believe humans are born in an "OK" state, that each person is unique, and that people ultimately are responsible for their own behavior and destiny.

Gilliland and James, 1998

HISTORY

Eric Berne (1910–1970), who was originally trained in the traditional psychoanalytic school, believed in "curing the patient." It was Berne's contention that the client should be included in the therapeutic process, thereby facilitating more proactive changes within the client (Woollams and Brown, 1978). The ideas of transactional analysis (TA) originated with Berne's work. He developed his theory of personality during the 1950s (Dusay and Dusay, 1989). Berne wrote simply and with humor, in a way that made the ideas of psychotherapy understandable and applicable for more people (Steiner, 1974).

Many of the ideas of TA are outlined in Berne's three major works: *Transactional Analysis in Psychotherapy* (1961), *Games People Play* (1964), and *What Do You Say After You Say Hello?* (1972). Berne's books, along with *Born to Win* (1971) by Muriel James and Dorothy Jongeward, and *I'm OK You're OK* (1973) by T. Harris, brought TA into the mainstream of North American culture in the early 1970s. Since that time, TA has become internationally known as one of the most respected and influential psychotherapies (Stewart, 1989). Within the school of TA, there has been much innovation and development. For example, out of Berne's original work, Robert and Mary Goulding developed *redecision therapy.*

For more information on TA write to: The International Transactional Analysis Association, 1772 Vallejo Street, San Francisco, CA 94123, or phone (415) 885-5992.

MAIN IDEAS

Dusay and Dusay (1989) assert that TA is an effective therapy because "the therapist and the client both share mutual tools and simple vocabularies, and while this tends to eliminate some of the therapist's magic, the client is facilitated to 'own' his appropriate share of responsibility for treatment."

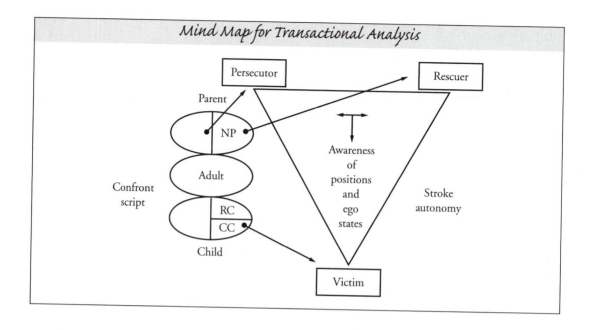

Ego States and Their Development

In his groundbreaking book, *Transactional Analysis in Psychotherapy* (1961), Berne describes *personality structure* as consisting of three basic ego state categories: Parent, Adult, and Child. Each ego state can be identified by distinct feelings, thoughts, and actions. Penfield's neurological research (Penfield and Roberts, 1959) tends to confirm Berne's three pragmatic absolutes (Figure 12–1). These are paraphrased as:

1. Every adult was once a child. (Child ego state)
2. Every adult is capable of reality testing. (Adult ego state)
3. Every adult has had parenting. (Parent ego state)

For transactional analysis (TA) to be useful to you as a human services counselor, you must become familiar with the ego states and how they transact with each other.

We were are all born with the Free Child ego state. The Free or Natural Child can be described as being the source of all our needs and the initiator of emotion and creativity (James and Jongward, 1971). Adaptations evolve as the Child ego state develops. These can be divided into two basic styles: the Compliant state and the Rebellious state (Woollams and Brown, 1978). The Compliant Child accepts the environment. The Rebellious Child rejects the environment.

The Parent ego state contains and expresses how we have learned to function or dysfunction as social beings (James and Jongeward, 1971). In other words, we internalize how we perceive the important people from our childhood. Beyond our mother and father, this might include a favorite aunt or an impressive teacher, depending on the experience of the child. The Parent ego state has two components: the Critical Parent and the Nurturing Parent (Woollams and Brown, 1978). The

Figure 12–1.

Functional diagram of ego states (adapted from Berne, 1961)

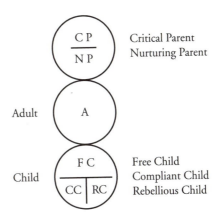

Critical Parent
Nurturing Parent

Adult

Free Child
Compliant Child
Rebellious Child

Child

Critical aspect is demanding and controlling, and the Nurturing component is comforting and supportive (James and Jongeward, 1971) .

Woollams and Brown (1979) describe the Adult ego state as "a probability-estimating computer." It is the part of us accommodating clear thinking and reality testing. It is straightforward and unemotional. The Adult ego state functions to meet the needs of the Child or Parent ego states.

Early Decisions

Transactional analysis asserts that the key to maintaining emotional healthiness is within the Child ego state (Goulding, 1987). Often the needs in an adult of the Free Child are encumbered by decisions made at a young age, under difficult circumstances, with the limited resources of a child (Steiner, 1974). Such decisions have the best intentions and are often guided by "magical thinking"; that is, the child believes she or he can affect the world more than is really possible (Stewart, 1989). For example, six-year-old Susan may think "If I stay away and play quietly, maybe Mommy and Daddy will stop fighting and everything will be all right." Perhaps Susan's decision is directly related to a parental *injunction* (order, or admonition) such as "You're bad!" or "Go away!" which she translates as *don't be close* (Goulding, 1987).

These early decisions contribute to habitual bad feelings or *racket* feelings that are connected directly to the early decision scene. Using the example of Susan: by staying with the early decision "don't be close," she distanced herself from her parents. Susan may have self-created depression as a racket feeling that helped to distance her from all relationships. These distancing maneuvers, which are not within her Adult awareness, are *games* that help fulfill the unconscious life plan of the client (Stewart, 1989). The early decision of "don't be close" may lead to the life plan, or *script,* "I will end my life alone" (Steiner, 1974).

Redecision Therapy

Redecision therapy is a set of theory and techniques that grew out of Berne's work with the incorporation of gestalt methods. Facilitated by a redecision therapist, Susan could follow the habitual feelings of depression back in time through various scenes of her life. The therapist encourages Susan to start with the present and go back to experience scenes that are congruent to the original feeling of depression. The therapist encourages the client to speak of past events as if they are occurring in the present. Each scene is explored and reexperienced vividly, through the senses and the emotions, until the scene is found in which the original decision was made (Woollams and Brown, 1978). The therapist asks "What are you thinking?" to discover the early decision as well as its underlying good intention (Goulding, 1987). The basic premise of redecision therapy lies in understanding that early decisions are based on the best intentions, considering the resources children have at their disposal at the time (Stewart, 1989). It is important for the therapist to acknowledge this whenever possible. Once the early decision situation has been explored, the client is invited back to the present, and the Adult ego state, to appreciate the good intention and to evaluate the current effect of the early decision (Goulding and Goulding, 1979).

If the client feels the early decision is now creating dysfunctional behavior, the therapist invites the client to "redecide" (Goulding, 1987). This might happen in a number of ways. Using our example of Susan, she may return to the original scene with her current Adult skills and resources, and simply say "I can be close to others. It is not my fault my parents are fighting."

When clients redecide, they are saying "I am not stupid, bad, crazy" and so on. The early decision is relinquished and the client experiences healthiness. The possibilities for change with redecision therapy are enormous, as clients give up a lifetime of obscuring painful feelings and experience the sense of well-being that is their right (Woollams and Brown, 1978).

Note: Redecision therapy should be practised only by those trained in its theory and practice, and supervised by trained professionals.

Strokes

One of the fundamental concepts, and basic interventions, in TA theory is the practice of understanding and applying *strokes* (James and Jongeward, 1971). The simplicity of strokes is often misunderstood. A *stroke* is defined as a unit of recognition (e.g., "When you did that for your mother, it showed how kind you are"). We need strokes, and we feel deprived if we don't get them (Stewart and Joines, 1987).

Berne (1961) teaches that we all have an innate desire he calls "stimulus hunger," which he defines as the need for physical and mental stimulation. Based on this idea, he recognized three types of strokes:

1. *Verbal and nonverbal.* As adults, we learn to accept other forms of recognition in conjunction with, or as a substitute for, physical touching, such as a smile, or a compliment. Berne (1961) used the term *recognition-hunger* to describe our need for this kind of acknowledgment by others.

2. *Conditional and unconditional.* A conditional stroke relates to what you do (e.g., "You are a great mother"). An unconditional stroke relates to what you are (e.g., "You have beautiful hair").

3. *Positive and negative.* A positive stroke is one the recipient experiences as pleasant. A negative stroke is experienced as painful. One might imagine people would

always seek positive strokes and avoid negative ones. However, in reality, any kind of stroke is better than no stroke at all (Stewart and Joines, 1987).

Stroke Economy \ Stroke City

Most of us live, to some degree, under the tyranny of *should* and other internal injunctions from the Critical Parent ego state. This is reflected in our culture, where positive strokes are often withheld or doled out in a miserly fashion. It is safe to say most of us could use more strokes (James and Jongeward, 1971). Eric Berne's student and collaborator, Claude Steiner, discussed this stroke deprivation as basic training in lovelessness, or what he refers to as the *stroke economy rules* (1974):

Stroke Economy Rules

- Don't give strokes (if you have them to give)
- Don't ask for strokes (when you need them)
- Don't accept strokes (if you want them)
- Don't reject strokes (when you don't what them)
- Don't give yourself strokes (no bragging)

The situation is simply reversed when individuals take responsibility for their own well-being. This individual is now said to follow *stroke city rules*:

Stroke City Rules

- Do give strokes (if you have them to give)
- Do ask for strokes (when you need them)
- Do accept strokes (when you want them)
- Do reject strokes (when you do not want them)
- Do give yourself strokes (brag)

Transactions

I have been discussing strokes in terms of our needs and emotions, which is the perspective of the Child ego state. When we look at strokes from the view of how we function, we are using the Adult ego state.

The exchange of strokes is referred to as *transactions*. More specifically, "A transaction is an exchange of strokes between two persons consisting of a stimulus and response between specific ego states. Transactions can be simple, involving only two ego states, or complex, involving three or four ego states" (Woollams and Brown, 1978).

Berne's Rules of Communication

Eric Berne devised three rules concerning communication and transactions.

Rule 1. As long as the transactions remain complimentary, communication may continue indefinitely (Figure 12–2). This rule covers the complementary or parallel transaction. The response comes from the same ego state the stimulus was directed to, and the response is directed back to the same ego state that

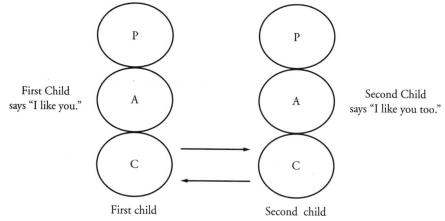

First Child
says "I like you."

Second Child
says "I like you too."

First child

Second child

Figure 12–2.

Complementary transaction: Two children in a sandbox

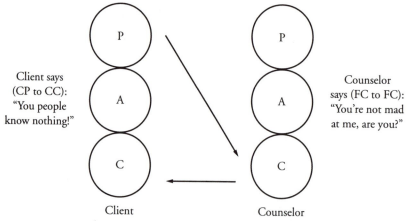

Client says
(CP to CC):
"You people
know nothing!"

Counselor
says (FC to FC):
"You're not mad
at me, are you?"

Client

Counselor

Figure 12–3.

Crossed transaction (see *options* in Techniques).

initiated the stimulus. Although a complementary transaction allows continuous communication, it does not necessarily indicate the content or value of that communication.

Rule 2. Whenever the transaction is crossed, a breakdown (sometimes only a brief, temporary one) in communication results, and something different is likely to follow. (Figure 12–3). When a response comes from an ego state other than the one indicated by the stimulus, it is said to be a crossed transaction.

Rule 3. The outcome of transactions will be determined on the psychological level rather than on the social level (Figure 12–4). In an ulterior transaction, there is an unspoken "secret" or psychological message beneath the spoken social stimulus message.

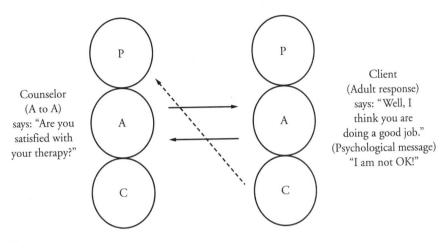

Figure 12–4.

Ulterior transaction (see Games).

Games

In TA, a *game* is a series of transactions between two or more people with an ulterior transaction that usually ends with a "payoff" of bad feelings (Woollams and Brown, 1979). Games are played to collect strokes, usually negative (Berne 1964). For example:

CLIENT: I do not know what to do! (Secret message: I'm not OK)

COUNSELOR: Would you like me to help you? (Secret message: Yes, you are not OK)

CLIENT: I doubt you can. (Secret message: You're not OK. Payoff: bad feelings for the counselor, negative strokes for both)

TECHNIQUES

1. *Contracting for change.* The therapeutic contract sets the focus for treatment. Clients decide specifically, in terms of beliefs, emotions, and behavior, what they plan to change about themselves in order to reach their designated goals. Clients work with the counselor to determine the contract and make the contract with self. The counselor serves as a witness and a facilitator to this process (Goulding and Goulding, 1979). Two questions that are often helpful in arriving at a contract are:

   COUNSELOR: What would you like to change today? (The emphasis is on the solution, not the problem, and the power to change is with the client.)

   COUNSELOR: When you succeed at making this change, how will your life be different? The answer to this question gives a clear picture of how the contract will be fulfilled.

2. *Ego states exercises.* Seeing and thinking in TA involves moment to moment awareness of both external and internal phenomena, and verbal and nonverbal

communication; therefore, experiential exercises can aid in learning about the ego states.

The different ego states are observable and recognizable by the client's postures and physiology. For example, finger shaking or a tight-lipped smile would indicate the Controlling Parent, as would use of language such as "you always," "you never," and "you should" (James and Jongeward, 1971). The posture of the Compliant Child would exhibit a sunken chest, with shoulders slumped, while the Rebellious Child's chest would be puffed out (James and Jongeward, 1971). The Nurturing Parent's tone of voice would be soothing, while the Adult's would be neutral (Woollams and Brown, 1978).

The first of these ego state exercises can be adjusted to suit the individual's memories of childhood. Create a visualization that rings true for you. If you have trouble remembering, imagining or pretending works well.

- Visualization Exercise
  Take some deep breaths . . . make yourself comfortable . . . close your eyes . . . let yourself return to a time when you were a very young child . . . as far back as you can remember . . . you are in the home of your childhood . . . look around you . . . you are in the living room or the kitchen . . . how are things arranged? . . . let your senses take in the room . . . the sounds . . . the smells . . . the colors . . . the furniture . . . the walls . . . enrich your visualization with the details that come to you . . . (pause) . . . as you are settling into the experience of the room, the important grownups in your life begin to enter one by one . . . you notice them . . . listen to what they are saying . . . . how they speak . . . how are they behave and act . . . take some time to relax into the scene, noticing the details . . . (pause) . . . now remember how you relate with the grownups . . . be with them . . . what do you do? . . . what do you say? . . . how do you feel? . . . stay with it . . . establish and enrich the scene . . . how are you responding to the grownups around you . . . (pause) . . . now remain with the visualization and listen to what I say about it . . . at your present age, when you talk, act, and feel like the grownups in your visualization, you are in your Parent ego state . . . when you talk, behave, and feel like the child, you are in your Child ego state. . .when you are observing events around you. . . thinking, gathering and sharing information . . . as if a part of you is now observing the scene . . . you are in your Adult ego state . . . when you are ready, come back into the here and now (adapted from Woollams and Brown, 1978).

You can gain further insight by discussing this exercise within your study group.

- Six Chair Technique (Figure 12–5)
  This exercise for developing awareness of the ego states (Porter Steele, 1996). has a gestalt flavor; each of the six chairs is the seat of a separate ego state.

  When initiating the six chairs, have clients bring to mind an issue around which there is a feeling of conflict. Begin with something easy. Have clients sit in the chair that corresponds to their current ego state. Us-

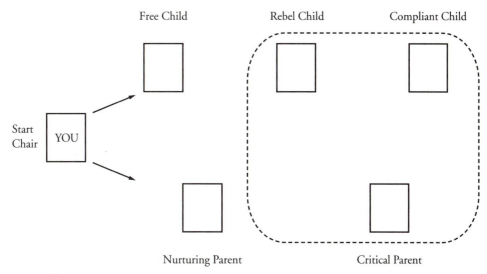

Figure 12–5.
Six-chair technique.

ing the issue, create a dialogue, moving from chair to chair with the changes in ego states. This may seem awkward, and it is helpful to have someone with experience coach you as you get used to the process.

If you are on your own, hang in there, your intuition will tell you how it is going. Remember, you are working with a model, and it is your phenomenological perspective that counts in this learning process. Not only are you learning the characteristics of the ego states, but also the styles of interaction among them. For example, in times of stress the personality is often dominated by the Critical Parent and the Compliant and Rebel Child ego states (in Figure 12–5, this is the area within the dotted line).

In stressful situations, the Free Child, the Adult, and the Nurturing Parent "can't get a word in edgewise." The counselor can intervene here with options that are helpful for the client. The counselor's job involves lowering the client's anxiety. Recognizing the ego states of the client, and knowing what transactions might be effective at the moment, will lower anxiety and foster the basic healthiness of the client.

3. *Stroke exercises.* These involve teaching and experiencing the dynamics of stroking. (All exercises in this section have been adapted from Stewart and Joines, 1987.)

- Find a partner. Walk towards each other. First, you take a turn greeting your partner. Your partner walks past, as if you were not there. Switch roles and share your experiences.

- In a group: Moving around the group, each person gives a positive conditional stroke to the person on his or her left. Notice each time how the stroke is given and how it is received. When the round is finished, discuss what you observed. Then go around in the other direction. Again discuss how the strokes were given and taken.

Questions: Of the strokes given, which were true and which were counterfeit? Did anybody "throw marshmallows" (e.g., you are so cute)? When people were taking strokes, who received the stroke with open appreciation? Who discounted the offered stroke? How did you see, feel, and hear them doing so? Did anyone openly refuse a stroke they did not want?

- Everyone take turns bragging (positive self-recognition) for one minute. Try to make your brags genuine, but don't be too serious. Spend some time discussing your experience.

4. *Options and the drama triangle.* Understanding the three types of transactions is basic to understanding the theoretical aspect of transactional analysis. Applying this theory with the experiential understanding of the ego states, the human services counselor can exercise what Karpman (1968) called *options* and guide communication in an effective manner. Karpman (1968) pointed out the many uses of crossed transactions in redirecting clients. For each stimulus directed by the client, the counselor can choose from five different ego states and innumerable responses. For example, a client may begin to escalate an attack by saying from Critical Parent, "You people know nothing!" directed at the Adapted (Compliant) Child of the counselor. The wealth of responses might include, from each counselors ego state:

- Adult to Adult. "You do have a grievance."
- Critical Parent to Adapted Child. "You have no idea what I go through to help you out!"
- Nurturing Parent to Adapted Child. "Hey, you kinda got lost in the shuffle around here."
- Rebel Child to Compliant Child. "Don't have a cow!"
- Free Child to Free Child. "You're not mad at me are you?" or "Isn't today the day you play basketball?"

Not all of these responses are necessarily going to be effective. Developing skill in applying options takes experience. Once you develop this skill, it can enrich counseling and crisis intervention considerably.

Another awareness tool from Karpman (1968) that can be used in conjunction with crossed transactions is the *game* or *drama triangle*. Using this, habitual activity and conflict can be seen to originate from three psychological positions: Victim, Persecutor, and Rescuer (Figure 12–6).

Awareness of clients' psychological positions in relation to the counselor and the transactions they make can help the counselor point out conflict and facilitate intimacy. Look back at the example from games, and you will notice the client moves from a victim position to the persecutor position. The following exemplify the three positions:

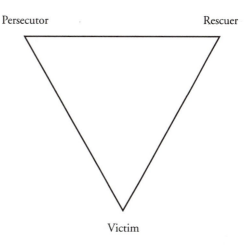

Figure 12–6.
Drama
triangle
(adapted from
Karpman,
1968)

PERSECUTOR: You always do it wrong!

RESCUER: I am only trying to help you.

VICTIM: Why is it always me?

5. *Confronting the "con."* In TA, recognizing the psychological or hidden message is part of the counseling process (see ulterior transactions, and games). When this message subtracts from the autonomy or "OKness" of the client, others, or the reality of the situation, it is seen as a *discount*. When a discount initiates a game, it is called a *con* (Goulding and Goulding, 1979). Confronting the con brings the discount into the Adult awareness of the client, and interrupts the game, facilitating autonomy and intimacy (Stewart, 1989). For example:

CLIENT: I never get what I want.

COUNSELOR: So, you never received anything you wanted in your whole life?
    A *gallows con* is presented as humor by the client, and it tends to increase the suffering of the client or "tighten the noose," hence the name:

CLIENT: Ha, ha, ha, what an idiot I am.

COUNSELOR: I do not see this as being funny; you appear anything but an idiot to me.

Summary

Transactional analysis is presented in this text because it is experiential and applicable for use by the human services counselor. Stewart (1989) teaches that, philosophically, TA can be summarized as the following:

- People are OK and fundamentally healthy.
- Everyone has the capacity to think.
- People decide their own destinies, and these decisions can be reexamined and changed.

Because of its down-to-earth language and experiential process, the accessibility of TA for the human services counselor may be its strongest selling point. It offers an explanation of how people experience varying degrees of difficulty in relating to others. I believe the techniques provided in this chapter are user friendly for counselors. Yet, a full understanding of the dynamics of transactional analysis and redecision therapy requires experience, instruction, and study.

I highly recommend that you have formal training and supervision before starting to use TA.

References

Berne, E. (1961). *Transactional analysis in psychotherapy.* New York: Grove Press.

Berne, E. (1964). *Games people play.* New York: Grove Press.

Berne, E. (1966). *Principles of group treatment.* New York: Oxford University Press.

Berne, E. (1972). *What do you say after you say hello?* New York: Grove Press.

Berne, E. (1975). *Intuition and ego states.* San Francisco: TA Press.

Dusay, J. (1977). *Egograms: How I see you and you see me.* New York: Harper & Row.

Dusay, J. (1981). Eric Berne: Contributions and limitations. *Transactional Analysis Journal 2:* 41–45.

Dusay, J., Dusay, K. (1989). Transactional analysis. In R. J. Corsini, D. Wedding (eds.), *Current psychotherapies* (4th ed.). Itasca, IL: F. E. Peacock.

Gilliland, B., James, R. (1998). *Theories and strategies in counseling and psychotherapy.* Needham Heights, MA: Allyn & Bacon.

Goulding, R. E., Goulding, M. (1978). *The power is in the patient: A TA/Gestalt approach to psychotherapy.* San Francisco: TA Press.

Goulding, R. E., Goulding, M. (1979). *Changing lives through redecision therapy.* New York: Brunner/Mazel.

Goulding, M. (1987). Transactional analysis and redecision therapy. In J. K. Zeig (ed.), *The evolution of psychotherapy.* New York: Brunner/Mazel.

Goulding, R. E. (1987). Group therapy: mainline or sideline? In J. K. Zeig (ed.), *The evolution of psychotherapy.* New York: Brunner/Mazel.

Harris, T. (1973). *I'm OK—You're OK: A practical guide to transactional analysis.* New York: Harper & Row.

James, M., Jongeward D. (1971). *Born to win.* Reading, MA: Addison-Wesley.

Karpman, S. (1968). Fairy tales and script drama analysis. *Transactional Analysis Bulletin* 7(26):39–43.

Klein, M. (1980). *Lives people live: A textbook of transactional analysis.* London: Wiley.

Penfield, W., Roberts, L.(1959). *Speech and brain mechanisms.* Princeton: Princeton University Press.

Porter Steele, N. Personal communication, March 20, 1996. Technique taught to Porter Steele by Barbara Hibner.

Steiner, C. (1974). *Scripts people live.* New York: Grove Press.

Stewart, I. (1989). *Transactional analysis counseling in action.* New York: Sage.

Stewart, I., Joines, V. (1987). *TA today.* Nottingham: Lifespace.

Woollams, S., Brown, M. (1978). *Transactional analysis: A modern and comprehensive text of TA theory and practice.* Dexter, MI: Huron Valley Institute Press.

Chapter 13

Gestalt Therapy: Fritz Perls

If you are in the now, you can't be anxious, because the excitement flows immediately into ongoing spontaneous activity. If you are in the now you are creative, you are inventive. If you have your senses ready, if you have your eyes and ears open, like every small child, you find a solution.

Fritz Perls, 1969

HISTORY

Gestalt therapy was developed by Frederick "Fritz" Perls (1893–1970) and his wife Laura Perls. Born in Germany at the turn of the last century, they began practising as psychoanalysts in the turmoil-filled time between the two World Wars. They started to practise gestalt therapy in Frankfurt in the 1940s. Perls was influenced by many other psychotherapists including Freud, Rank, Reich, Adler, and Jung. Fritz Perls found that psychoanalysis, with its digging into the past and blaming others, did not seem to be helping many of his clients. The Perlses started out by practicing therapy in Amsterdam, moved to South Africa in 1934, and later went to Montreal. In 1941, the Perlses established the New York Institute for Gestalt Therapy, where they practised and trained others. They retired on the west coast of Canada in 1970.

For more information, write to Gestalt Institute of Cleveland, 1588 Hazel Drive, Cleveland, OH 44106.

MAIN IDEAS

We have included Fritz Perls's gestalt concept to highlight the importance of awareness of environment in the counseling process. Simply helping clients become more aware of what they are doing in an environment can bring insight and facilitate change. Fritz Perls (1969) wrote:

> So it's a question of being, rather than having (an existence). This is why we call our approach an existential approach: we exist as an organism—as an organism like a clam, like an animal, and so on, and we relate to our world just like any other organism of nature . . . wherever we go we take a kind of world with us.

Everything we do, think, believe, and remember affects our whole being. An organism's primary inborn motives are toward *self-preservation* and *actualization of the self* (Perls, 1969). Perls believed that we all attempt to *actualize* ourselves (defined as

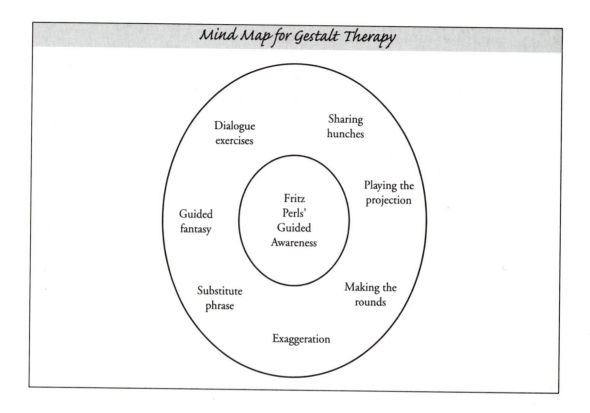

Mind Map for Gestalt Therapy

Dialogue exercises

Sharing hunches

Playing the projection

Guided fantasy

Fritz Perls' Guided Awareness

Substitute phrase

Making the rounds

Exaggeration

"the basic human drive toward growth, completeness, and fulfillment") (Yontef and Simkin, 1989). In gestalt therapy, people are considered to have the inner wisdom to change for their own well-being (Gladding, 1988).

Awareness

In gestalt therapy, awareness is the main goal and the primary tool (Yontef and Simkin, 1989). Because gestalt therapy is based on this premise, we have chosen techniques that emphasize the development of awareness. Once awareness is achieved, excitement develops; that is, energy emerges from within the person. When clients become more aware of the present moment, they are in a position to own their individual choices. Through contact with the moment, the client will be able to begin natural and spontaneous change (Perls, 1969). Contact is said to be the lifeblood of growth; change is inevitable when we make contact with others and the natural world without losing our sense of individuality. The most valuable attribute the therapist has is the ability to relate to the client's area of concern and to assist the client in increasing awareness (Bernard, 1986).

Resistance to Awareness or Contact

Anything that impairs awareness is seen as resistance to awareness, or as a barrier to contact (Perls, 1969). When our upbringing teaches us to regulate ourselves by external rules and to ignore our insight, we give up our internal locus of control, and we give up our present choice of freedom (see Chapter 15). The goal of awareness is to have the client take responsibility and control for self (Todd and Bohart, 1994). This is done by allowing the client the opportunity to come into contact with the authentic self. A famous Perls statement was that "getting to the core of personality is much the same as peeling off all the individual layers of an onion." Five layers of Perls's onion included:

1. *Phony layer*—reacting to others in stereotypical and inauthentic ways.
2. *Phobic layer*— avoiding the emotional pain associated with seeing ourselves the way we really are.
3. *Impasse layer*—the point where we are stuck in our own maturation.
4. *Implosive layer*—exposing our defenses and beginning to make contact with our authentic selves.
5. *Explosive layer*—releasing an an enormous amount of energy as we let go of phony roles and pretenses.

The counselor will, through various techniques, assist the client at the impasse level to face internal resistance to becoming authentic.

There are five defenses that result in impasses (Polster and Polster, 1973):

1. *Introjection*—uncritically accepting others' beliefs and standards without assimilating them to make them congruent with who we are (e.g., the youth who is defined as a follower).
2. *Projection*—the reverse of introjection. Disowning certain aspects of ourselves (the attributes of our personalities that are inconsistent with our self-image), and put them onto other people (e.g., the adult who never gets off the couch, but criticizes the children for being lazy).
3. *Retroflection*—turning back onto ourselves what we would like to do to someone else, or doing to ourselves what we would like someone else to do to us. For instance, directing anger inward that we are fearful of directing at others (e.g., a youth considers suicide because the parents split up).
4. *Deflection*—diffusing authentic contact through the overuse of humor, abstract generalizations, and questions rather than statements (e.g., the class clown, who is always joking but finds school difficult and is academically unsuccessful).
5. *Confluence*—a blurring of the differentiation between self and others. Going along with others and not expressing true feelings and opinions. A belief that all persons experience the same feelings and thoughts (e.g., the wife who irrationally refuses to abandon an abusive relationship).

Experiments

I encourage the human services counselor to view counseling sessions as a series of *experiments* designed to allow the client to gain awareness in an experiential way. However, I want to caution you never to have an experiment that revictimizes the client. The counselor needs to ensure that the client is aware of the potential gains, as well as pains, that any experiment may provide. The counseling session is a creative adventure, in which what is learned can be a surprise to both the therapist and the client. Gestalt therapy experiments facilitate the client's ability to work through impasses (stuck points) in personal growth. Experiments encourage spontaneity and inventiveness by bringing possibilities for action directly into the therapy session (Latner, 1973). They bring conflict to life by having the client experience and enact the scene in the first person, present tense. Clients *experience* the feelings associated with their conflicts, as opposed to talking about them. As Polster (1987) explains, through experiments clients actually experience the feelings associated with conflicts that are past, present, or future, imagined or real.

Experiments can take many forms, from which Corey (1996) provides the following suggestions:

- Imagining a threatening future encounter
- Setting up a dialogue between the client and some significant person
- Dramatizing the memory of a painful event
- Reliving in the present a particularly profound early experience
- Assuming the identity of one's mother or father through role playing
- Focusing on gestures, posture, and other nonverbal signs of inner expression
- Carrying on a dialogue between conflicting aspects within oneself

It is important to realize how strange some of these experiments may seem to the client. To utilize them effectively, counselors need to experience personally the power of gestalt experiments. Trust in the client/therapist relationship is the key to client participation. Again, it is important not to start experiments that are beyond your training or certification.

TECHNIQUES

The techniques included in this chapter have been chosen from an array of gestalt methods and focus on the ways in which the human services counselor can help a client become aware of all aspects (physical, psychological, and social) of the client's behavior in the "here and now."

Gestalt therapy has the dynamic quality of the gestalt process, which helps the client become responsible for self-supportive and congruent behavior (Todd and Bohart, 1994). Always bear in mind Perls's caution that this kind of counseling "takes a lot of concentration, but so does almost anything that is worth doing . . . and Gestalt Therapy is not a one-track thing" (1976).

1. *The dialogue exercise.* One goal of gestalt therapy is to promote integration and acceptance of aspects of the self that have been disowned and denied. This is accomplished through a form of role play in which the client plays all the parts that is also known as *empty-chair* or *two-chair* work. It is used to help clients become more aware of both sides of a conflict, whether the conflict is within the self or with another person (Todd and Bohart, 1994). Internal conflicts are rooted in introjection; it is essential to become aware of these introjects, especially the toxic introjects that poison the system and prevent personality integration. Chair work is one way of helping the client to externalize the introject.

There are a variety of conflicts that can be addressed through chair work. When taking on roles, it is most effective when the client speaks in the first person, present tense, as though the dialogue is actually taking place in the present. For example:

CLIENT: [Talks about a conflict with mother.]

COUNSELOR: [Asks client to pretend that mother is sitting in the empty chair and uses dialogue between them to work out the conflict from a phenomenological perspective.]

CLIENT: My mother never loved me! I really wish we could have worked things out before she died.

COUNSELOR: Why don't you pretend she's sitting right there in that chair and tell her how you feel.

CLIENT: OK. Mom, why didn't you ever tell me you loved me.

COUNSELOR: Now move to the other chair and answer as your mother.

CLIENT (AS MOTHER): I tried to tell you. I just didn't know how to. I wish that I had told you how much I really loved you. You are my daughter, and you have to know that I do love you.

This dialogue would be continued until some form of resolution occurred. Potentially the scene could open up to include other characters such as siblings, the father, or other significant persons, fictional or real (Polster, M., 1987). This technique can result in a re-invention or re-decision, empowering the more healthy and autonomous parts of the self.

2. *Sharing hunches.* The counselor or group members share feelings and perceptions of other members in a tentative manner or in the form of an intuition or hunch; this provides clients with more insight as to how others see them. Permission must be requested from the client before giving each hunch. For example:

COUNSELOR: I have a hunch about your situation, would you like to hear it?

CLIENT: Sure.

COUNSELOR: My hunch is that you are feeling unworthy and unaccepted at home.

3. *Substitute phrase.* This demonstrates and helps make clients aware that they are splitting off certain parts of their conscious experience to avoid acting responsibly (Todd and Bohart, 1994). This is an effective form of positive self-talk. For example:

CLIENT: I can't confront that person. [Or, "I can't control my anger."]

COUNSELOR: Please repeat that statement using the words *I can.*

4. *Exaggeration.* The counselor asks the client to exaggerate a certain body movement, feeling or thought that has just manifested, to intensify the experience

and bring it into present awareness (Yontef and Simkin, 1989). It also allows the client to become more aware of personal defenses. Blocked energy is a form of resistance and can be manifested physiologically. The counselor encourages expression and awareness of this blocked energy in order to facilitate resolution. For example:

- The bereaved client, who is showing little emotion, is asked to exaggerate a small sigh that has been made. The client sighs deeply twice and begins to murmur softly. The counselor asks the client to give words to the sighing; the client's feelings surface and are expressed. This assists the client to come into contact with the moment and become more aware of emotions.

5. *Making the rounds.* This exercise involves having members of the group go up to each other in the group, and make contact. This helps clients to learn to take risks, to confront others, to be spontaneous, to express their needs, and to experiment with new behaviors. It creates an environment of immediacy (Latner, 1973). For example:

- Clients may ask each member of the group to give them a statement of positive recognition. If a group member states that he or she does not feel safe in the group situation, the leader may encourage that member to approach each person in the group during breaks and finish a sentence like the following: "I don't feel safe here with you, because . . . " (e.g., I don't feel safe here with you because I have never been in a group before).

6. *Guided fantasy.* This is another awareness-building tool that is designed to show clients how they create their own anxieties (Yontef and Simkin, 1989). The purpose is to teach the clients to become aware "in the moment" of how their thinking affects them. For example:

- A male client is asked to imagine being with a co-worker whom he likes, but whose unacceptable behaviors he finds impossible to confront. The client describes his thoughts and feelings from movement to movement in the imaginary scene. It is important that the client speaks in the first person, present tense (so that he is "doing," rather than "talking about"). At this point, the client is in a position to learn a new skill that can be utilized in the real conflict. The counselor offers the client a substitute phrase to replace his paradigm so he is able to confront this unacceptable behavior. At this point, the counselor guides the client through the situation once again to obtain a more positive outcome.

7. *Playing the projection.* Statements that clients make about others are often projections of attributes that they possess themselves. To help bring awareness of inner conflicts, the counselor asks the client to play the role of the recipient of the statements he or she makes about others (Corey, 1996). For example:

- The female client states that she hates how a certain person gossips about others. The counselor asks her to play the role of the other person. This can facilitate her discovering to what degree the statement represents her own inner conflicts.

Summary

Gestalt therapy helps clients to become aware of their present behavior in preparation for learning or choosing new behaviors. It is based on a philosophical concept that makes a strong distinction between what one actually knows and what one perceives through subjective, often fallible, senses. Gestalt therapy, like other phenomenological orientations, focuses on the present moment, which allows the client, through experiments, to increase awareness of personal experience. Having present experience brought into awareness creates insight for clients, leading them to see significant factors that affect their whole being.

Gestalt therapy is a holistic theory, claiming that a person's behavior is part of a field of being, as opposed to distinct sequences of cause-and-effect. Taking a client momentarily out of the subjective world and into an awareness of elements in the here-and-now is generally acknowledged to be of therapeutic benefit (Polster, E., 1987).

Note: Once again I caution you to ensure that you have been professionally trained and supervised before attempting any gestalt experiments that may be outside the realm of safe practice. The techniques provided in this section are geared for the beginning counselor; however, they have also been utilized by more experienced counselors.

References

Bernard, J. M. (1986). Laura Perls: From figure to ground. *Journal of Counseling and Development* 64:367–73.

Corey, G. (1995). *Theory and practice of group counseling* (4th ed.). Pacific Grove, CA: Brooks/Cole.

Corey, G. (1996). *Theory and practice of counseling and psychotherapy* (5th ed.). Pacific Grove, CA: Brooks/Cole.

Gladding, S. T. (1988). *Counseling: A comprehensive profession.* Columbus, OH: Merrill.

Greenberg, L. S. (1980). Training counsellors in Gestalt methods. *The Canadian Counsellor* 15:174–80.

Hunt, M. (1993). *The story of psychology.* New York: Doubleday.

Kottler, J. (1994). *Advance group leadership.* Pacific Grove, CA: Brooks/Cole.

Latner, J. (1973). *The Gestalt therapy book.* New York: Bantam.

Perls, F. S. (1969). *Gestalt therapy verbatim.* Lafayette, CA: Real People Press.

Perls, F. S. (1976). *The Gestalt approach.* New York: Bantam.

Perls, L. (1976). Comments on the new directions. In E. Smith (ed.), *The growing edge of Gestalt therapy.* New York: Brunner/Mazel.

Pfeiffer J. W., Pfeiffer J. A. (1975). A Gestalt primer. In J. W. Pfeiffer, J. E. Jones (eds.), *The 1975 Annual Handbook for Group Facilitators.* San Diego: University Associates.

Polster, E. (1987). Escape from the present: Transition and storyline. In J. K. Zeig (ed.), *The evolution of psychotherapy.* New York: Brunner/Mazel.

Polster, M. (1987). Gestalt therapy: Evolution and application. In J. K. Zeig (ed.), *The evolution of psychotherapy.* New York: Brunner/Mazel.

Polster, E., Polster, M. (1973). *Gestalt therapy integrated: Contours of theory and practice.* New York: Random House.

Todd J., Bohart, A. C. (1994). *Foundations of clinical counseling and psychology* (2nd ed.). New York: HarperCollins.

Yontef, G. (1982). Gestalt therapy: Its inheritance from Gestalt psychology. *Gestalt Theory 4:* 23–39.

Yontef, G., Simkin, J. (1989). Gestalt therapy. In R. J. Corsini, D. Wedding (eds.), *Current psychotherapies.* Itasca, IL: F. E. Peacock.

Chapter 14

More Directive   Less Directive

Neurolinguistic Programming:
Richard Bandler and John Grinder

The meaning of your communication is the response that you get. If you can notice that you are not getting what you want, change what you're doing.

Bandler and Grinder, 1979

HISTORY

Neurolinguistic programming (NLP) is a relative newcomer to the world of psychology, having been developed in the early to mid 1970s by John Grinder, a professor at the University of California at Santa Cruz, and Richard Bandler, one of his graduate students (Robbins, 1994).

Bandler and Grinder, a linguistic specialist, combined their talents to explore human behavior using a behavioral-psychology general systems theory to develop a model for effectively communicating with clients in order to meet their needs. Bandler and Grinder (1975) termed their methodology *human modeling,* the study of how people perform or accomplish tasks.

Neurolinguistic programming is actually a collection of techniques Bandler and Grinder found useful in helping clients become more attuned to their internal or subjective experiences (Jacobsen, 1994). Neurolinguistic programming techniques come from observations, as well as from adaptations of the work of Virginia Satir, Milton Erickson, Fritz Perls, and Gregory Bateson (Harman and O'Neil, 1981). I believe that human services counselors can benefit from the technology of NLP.

MAIN IDEAS

Neurolinguistic programming is included in this text to help the professional improve skills in the area of communication. Many of us, when we think of communication, consider only its verbal component; however, as Adler and Towne (1993) explain, nonverbal communication is 12.5 times more powerful than words. Bandler and Grindler (1975) believe that, in order to understand language and communication fully, both verbal and nonverbal language need to be observed.

Grinder and Bandler (1976) teach that everyone's perceptions of the world (what they believe is happening at any time, at a conscious level), are filtered through

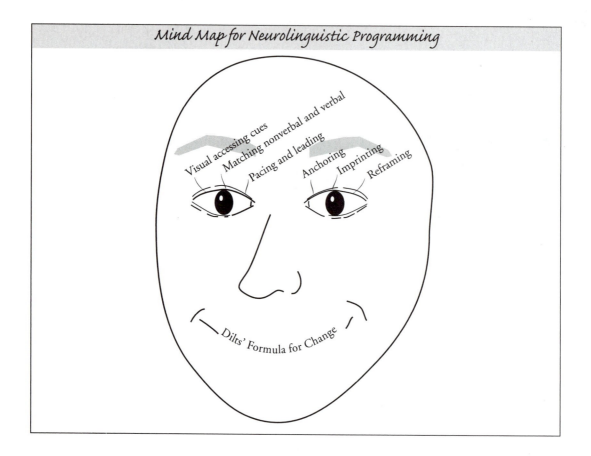

Mind Map for Neurolinguistic Programming

Visual accessing cues
Matching nonverbal and verbal
Pacing and leading
Anchoring
Imprinting
Reframing
Dilts' Formula for Change

their own unique sensory systems. No two people will ever perceive the world the same way (Bandler and Grinder, 1979). James (1997) explains that perception is projection and, to help clients, we need to assist them to put themselves at *cause* (operating from an internal locus of control). Thus, multiple experiences of external reality can never be the same; each individual will interpret external stimuli internally and formulate a personal interpretation of what is actually happening (Marvell-Mell, 1989).

All external events are first processed by the brain at an unconscious level, then internalized at a conscious level; from this, people create and generate their external behaviors (Bandler and Grinder, 1975). From their research, Grinder and Bandler (1982) concluded that, by recognizing and understanding language patterns in the client, the counselor can help a client process and adjust interpretations of the external world. Marvell-Mell (1989) stated "the sensory receptors and language act as filters for human experience. Communication difficulties often arise when an individual is limited from making full use of either his/her sensory facilities or his/her linguistic representations of experiences."

Bandler and Grinder (1979) use the term *modality* to refer to one of the five senses. They further teach that the components of a modality represent *sub-*

modalities (e.g., in the auditory modality, sub-modalities include volume, rate, pitch) (1982). Dilts and Megers-Anderson (1980) explain that neurolinguistic programming studies the visual, auditory, and kinesthetic sensory facilities. Neurolinguistic programming does not recognize the other two senses, smell and taste, as having any benefit in gathering and understanding communication and language (Marvell-Mell, 1989).

Understanding the role of sensory faculties enables the counselor to recognize that people usually develop one of the three sensory systems as their primary way of communicating, learning, and processing the external world (Dilts, Hallbom, and Smith, 1990). However, it is important to remember that we use all three systems on some level. Bandler and Grinder (1979) point out that everyone has the capacity to learn how to improve and develop under-utilized systems. They also indicate that it is rare for anyone, without being trained or educated, to use all three primary systems equally.

Dilts and colleagues (1990) use the term *representational system* to define the process of taking in internal and external information (pictures, sounds, feelings) as a way of "representing" and defining the world. In NLP, these representational systems are often referred to as *primary modalities.* For example, Bandler and Grinder (1975) explain that people who rely on the visual modality may run movies in their head when they try to process information, an auditory individual may replay information like a tape recorder, and a kinesthetic individual may attach a body sensation to a particular experience. When people process external stimuli, they may use several different representational systems to obtain internal recognition; the favored process is known as the *lead system.* When accessing information, a person may begin with the visual, auditory, or kinesthetic representational system (Bandler and Grinder, 1975). For example:

- You are asked, "What was the happiest day of your life?" (Figure 14–1).
- First, you may access your visual memory (V^r). These are mental pictures stored in your memory that are associated with specific objects and situations. You may then go to your kinesthetic (K) memory. Here you have stored tactile and affective inputs. Finally, you may access the auditory internal dialogue (A^i). This contains the specific details associated with the event.
- Having gone through your lead system, you may answer "when my daughter was born."

It is helpful for the professional to learn clients' main modalities (primarily visual? kinesthetic? auditory?) as well as their lead systems (how they process through the different modalities). In the preceding example, the person's lead system was $V^r + K + A^i$. Bandler and Grinder (1979) teach "if you can determine what a person's lead and representational systems [modalities] are, you can package information in a way that is irresistible for him." An example of a question for a client with the above lead system might be: "Can you see a time in the past when you felt great and are you aware of all the good things that were present?"

Neurolinguistic programming teaches that not all clients communicate or learn in the same manner. By becoming aware of the client's major representational system, the human services counselor can create an orientation that will fit the client's

Figure 14–1.
Visual accessing cues for a normally organized righthander (adapted from Bandler and Grinder, 1979)

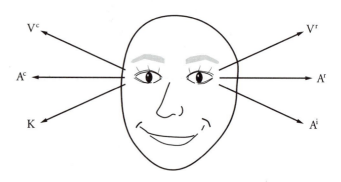

V^c = Visual constructed words.

V^r = Visual remembered images.

A^c = Auditory constructed sounds and words.

A^r = Auditory remembered sounds and words.

K = Kinesthetic feelings.

A^i = Auditory internal dialogue.

preferred modality of learning. It is important, when a particular counseling intervention is not working, to stop and try something different. *Resistance is not a client problem; it is a counselor problem.* In other words, the counselor has to ascertain the client's modality and find an intervention the client can process. When this is accomplished, resistance will likely vanish. For example, a visual client may not benefit optimally if counseling is entirely based on talking. If the client's main representational system (as in the example) is visual, the counselor with an eclectic orientation utilizes visual techniques (white board, pictures, graphs). When the professional is able to provide information through the dominant modality of the client, information can be processed efficiently in a language that has meaning to the client.

TECHNIQUES

The techniques we have chosen are meant to provide the human services counselor with a basic introduction to neurolinguistic programming techniques. This section is by no means a complete representation of neurolinguistic programming.

1. *Observing visual accessing cues (nonverbal communications).* By using visual accessing cues, the counselor may be able to formulate an early hypothesis as to how a client processes particular information and which representational system and lead system are preferred. Conyers (1995) claims that the eyes are the gateway to the human brain, both conscious and unconscious. Dilts and colleagues (1990) explain that the counselor can gain insight as to which modality the client is using if, for the first few seconds after asking a question, the counselor focuses not on what the client says but what the client does. Figure 14–1 can be

seen as a map that explains the position and meaning of various eye cues (*note:* the accessing cues in this figure are for righthanded persons, and it is not unusual for a lefthanded person to have reverse cues). The accessing cues provide the counselor with a nonverbal message regarding the representation and lead systems the client is using to process external stimuli. For example:

- Marvell-Mell (1989) provides an excellent example. When asking a client "What is bothering you today?" or "What do you want?", if the client moves the eyes up to V^r position and then responds "I have no idea," this may suggest that the information is present but out of the client's conscious level of awareness. Knowledge of the client's main learning modality provides the professional with a valuable insight that may be used to assist the client in accessing the stored information. For example, the counselor may ask a visual learner "How would you like to see things?" Bandler and Grinder (1975) suggest that some questions may not have a reference file for the client to access at an unconscious level (e.g., "What would it look like to have peace in your life?"). At this point, the professional should assist the client in developing a reference system for what peace would look like, using her eclectic questions in modeling the client's main representational system.

2. *Matching nonverbal and verbal behaviors.* In order for counselors to be able to match and model nonverbal and verbal behavior, they must be aware of *calibration.* Grinder and Bandler (1976) explain calibration as a process in which the counselor uses the three sensory modalities to be aware of external shifts in the client (changes in body posture, facial expression, breathing, skin color). This allows the counselor to be in tune with internal changes that are occurring. When counselors are calibrated, they become aware of their clients' processing loops for external stimuli. In other words, the counselor will be able to predict the lead system the client is going to use to obtain internal understanding and the behavior that will follow (quality visual images for a visual person will assist to relax their physiology).

   Pacing and leading. Pacing and leading are tools used by counselors to duplicate a client's behavior. They work at the client's unconscious level of awareness and are used to help keep the client's behavior and interpretations of the external stimuli from becoming overwhelming and unmanageable (Marvell-Mell, 1990). In fact, one of the most powerful tools for building rapport with the client is matching the client's behavior. For example:

CLIENT: [Begins to tell counselor about a concern.]

COUNSELOR: [As the client tells the story, the counselor starts physically to match or mirror the client's behavior. *Mirroring* is done as if the client were looking in a mirror (e.g., if the client moves his right hand, the counselor moves his left hand, mirroring the client's movements); *matching* is moving the same body part as that of the client.]

CLIENT: [The client starts to escalate behaviors (e.g., raising voice) during the discussion.]

COUNSELOR: [The counselor stops mirroring (pacing the client) and starts *leading* the client by changing voice and creating new postures.]

CLIENT: [The client starts to pace the counselor (unconsciously) and starts to relax (breathing slows down, voice calms).]

COUNSELOR: [The counselor continues to lead until the client stops pacing. At this point, the counselor resumes pacing the client until there is another escalation of behavior. The process repeats.]

Verbal matching. Neurolinguistic programming teaches that a counselor can help a client faster and more effectively to interpret internal and external information by working from the same modality; this is termed "modeling the client's representation systems" (Bandler and Grinder, 1979). Examples are:

Kinesthetic Matching

CLIENT: I feel I have no idea what I want in life.

COUNSELOR: When was the last time you felt you knew what you wanted in life?

Visual Matching

CLIENT: I see the future as being dark.

COUNSELOR: When was the last time you saw any light?

Auditory Matching

CLIENT: I only hear how scary the future is.

COUNSELOR: Have you ever heard of a not-so-scary future?

3. *Anchoring.* Bandler and Grindler (1975) explain anchoring as a natural process in NLP counseling. To anchor the client, a counselor intentionally sets up a stimulus (e.g., gestures). Anchoring the client begins by pairing an external stimuli, set up by the counselor, with an internal response in the client, in order to create a stable state.

Anchoring is similar to calibrating and becoming aware of lead systems. All of these techniques allow the counselor to formulate a prediction of how the client will process information, and what the client's behavior will be. Not until after counselors are aware of their clients' primary representational system and lead system are they in a position to calibrate the responses of the client, which then allows anchoring. Clients can be anchored using any of the representational systems; the key is to find an anchor that comes from their main representational system so they can quickly access and understand it. For example:

CLIENT: [Starts to talk about an issue that is beginning to cause emotional upset (e.g., unfinished business with father).]

COUNSELOR: [From an earlier discussion, the counselor remembers the client telling about playing in the park with the father. The client provided a clear visual explanation. When client talked about this memory, there was more color in the face, relaxed breathing, and a statement of seeing the self as happy.]

The counselor asks the client to go back to the park, picture playing with the father, and recall the wonderful memories. Shortly after the client starts to process this request, without being consciously aware, the client becomes more relaxed and less emotionally upset. At this point, the counselor paces the client through the issue of concern, using the memory to act as a safe guide. By asking the client to go to a "safe spot," the counselor has installed a visual /auditory anchor. The more it is used, the more it will strengthen.

The key to anchoring is for the counselor to wait for a physiological change that shows the client to be anchored and thus in a stronger position to address the concern. Note that anchors are easier to establish when working in the clients primary representational system.

4. *Imprinting.* Dilts and colleagues (1990) postulated that, once a person knows the desired outcome and has a clear understanding of what the self desires, imprinting may be initiated. This predetermined goal (desire) will be planted in a person's mind at an unconscious level of awareness. This increases the likelihood of achieving the goal. The term *cybernetic mechanism* describes this process in the brain (Dilts et al., 1990). For example:

CLIENT: Now I have a clear picture of what it looks like to be happy.

COUNSELOR: I have a process I believe will help you keep this picture. Would you like to look at it?

CLIENT: Sure!

COUNSELOR: [After explaining the process of imprinting, suggests the client look at reenforcing this picture ten times a day for the next week, and together they will observe the results.]

CLIENT: Great, ten times a day I will picture myself living a happy life.

5. *Reframing.* Many of us have walked into situations and encountered negative stimuli that led us to formulate negative internal interpretations. These situations worked on much the same premise as anchors. Automatically, they bring about a particular representational system and state (e.g., you are sensitive about your job performance and any time someone points out a mistake you become upset). In reframing, the individual learns to recognize negative thoughts and statements and replace them with positive ones (Grinder and Bandler, 1982). For example:

CLIENT: [before reframing] I am dumb because I made a mistake.

CLIENT: [after reframing] I am brave for attempting something new.

6. *Dilts' formula for change.* Human services counselors need to expand their skills and understanding of NLP before they can fully appreciate how powerful this formula is (Dilts et al., 1990). Once the counselor has helped the client self-evaluate the present state and formulate the desired outcome of counseling, the client and the counselor can work together to explore the client's resources and what must be learned to achieve the desired emotional state. The counselor also explores the limiting beliefs that may be preventing the client from obtaining

Figure 14–2.
Formula for
change
(adapted from
Dilts et al.,
1990)

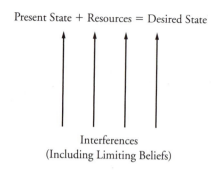

Present State + Resources = Desired State

Interferences
(Including Limiting Beliefs)

the necessary skills and information for goal achievement. A powerful question suggested by Dilts is *"What will you have to give up to achieve the desired state?"* Figure 14–2 presents a visual diagram to explain this concept. Many clients are experiencing *secondary gain,* meaning they are benefitting from undesirable behaviors. Some clients know the interfering behaviors are preventing the achievement of their goals; however, because of secondary gain they choose not to give up old behaviors. The client needs to make the choice, and the counselor needs to remember that the client is at cause; that is, the only one who can make the choice to stop the interfering behaviors. For example:

CLIENT: I can't see myself losing twenty pounds even though I really want to.

COUNSELOR: When we talk about what you want, you describe a picture of being fit. I would like to ask what you see yourself having to give up to get what you want.

CLIENT: Eating when I am depressed. When I see myself losing control, the only way I can find to regain control is through eating. Yet I see this is wrong for me.

COUNSELOR: [The counselor works with the client to explore the potential resources that will provide the same sense of control that food does.]

Until the client tries new behavior, most likely things will not change. The counselor has to help the client develop a picture of what is desired and what must be done to achieve it.

Summary

Neurolinguistic programming is being used in the fields of counseling, education, and business. I believe NLP helps counselors communicate effectively with clients. In the professional field, nothing outweighs the importance of good communication. We need to learn to communicate in a manner the client understands. Bandler and Grinder (1979) agree that NLP fits the model of the client, rather than having the client fit into the counselor's model.

As a human services counselor, you will find NLP an excellent resource for building rapport. There is extensive training available in this powerful counseling model. I have provided only a few of the introductory techniques that may be useful for your counseling orientation.

References

Adler, R., Towne, W. (1993). *Looking out/looking in.* Orlando: Harcourt Brace.

Bandler, R., Grinder, J. (1975). *Structure of magic I.* Palo Alto: Science and Behavior.

Bandler, R., Grinder, J. (1979). *Frogs into princes.* Moab, UT: Real People Press.

Conyers, M. (1995). *Supersuasion program.* Portland, ME: International Learning Institute for Accelerated Development.

Dilts, R., Grinder, J., Bandler, R., Delozier, J., Cameron-Bandler, L. (1979). *Neurolinguistic programming I.* Portland, OR: Metamorphous Press.

Dilts, R., Hallbom, T., Smith, S.(1990). *Beliefs pathways to health and well-being.* Portland, OR: Metamorphous Press.

Dilts, R., Megers-Anderson (1980). *Neurolinguistic programming in education.* Santa Cruz, CA: Not Ltd.

Grinder, J., Bandler, R.(1976). *The structure of magic II.* Palo Alto: Science and Behavior.

Grinder, J., Bandler, R.(1982). *Reframing: Neurolinguistic programming and the transformation of meaning.* Moab, UT: Real People Press.

Harmon, R., O'Neil, C. (1981). Neurolinguistic programming for counselors. *Personnel and Guidance Journal 59:*449–53.

Jacobsen, S. (1994). *Neurolinguistic programming.* INFO-LINE, American Society For Training and Development, April.

James, T. (1997). *Master practitioner, neurolinguistic programming.* Presented at Advance Dynamics Neurolinguistic Training, Kona, HA.

Marvell-Mell, L. (1989). *Basic techniques, book I.* Portland, OR: Metamorphous Press.

Robbins, S. (1994). *Neurolinguistic programming.* INFO-LINE, American Society For Training and Development, April.

Torres, C. (1986). The language system diagnostic instruction. In J. W. Pfeiffer, L. D. Goodstien (eds.), *The 1986 Annual: Developing human relations.* San Diego: University Associates.

Chapter 15

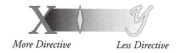

More Directive   Less Directive

Choice Theory and Reality Therapy: William Glasser

A therapy that leads all patients toward reality, toward grappling successfully with tangible assets of the real work, might accurately be called a therapy toward reality, or simply reality therapy.

William Glasser, 1965

HISTORY

William Glasser was born in Columbus, Ohio, in 1925. Hobart Mowrer wrote in the foreword of Glasser's 1965 *Theory and Practice of Group Counseling:* "Late in his training as a psychiatric resident, Glasser saw the futility of classical psychoanalytic procedures and began to experiment with a very different therapeutic approach, which he eventually named reality therapy." Glasser (1984) later developed, with the help of William Powers, what he termed *control theory* to support the techniques of reality therapy.

In his book *Choice Theory: A New Psychology of Personal Freedom* (1998), Glasser updated his original work on control theory and changed the name to *choice theory*. Glasser has not only had an influence in the field of psychology through his development of choice theory and reality therapy but he has also been a major influence on education through his book *Quality Schools*. In addition, the business world has benefitted from *Control Theory Manager* (1994). His most recent text, *Staying Together* (1995), focuses on marriage relationships.

Today, Glasser is president of The Institute for Choice Theory, Reality Therapy, and Quality Management. For more information, write: 22024 Lassen Street, #118, Chatsworth, California 91311.

MAIN IDEAS

Many counselors enjoy using reality therapy because it is user-friendly, has simple terminology, employs logical progression, and provides an effective questioning technique easily understood by both client and counselor. Reality therapy is a philosophy of life that has extensive application to nearly every aspect of human relationships. William Glasser (1984) identifies the most fundamental needs that drive us as love and belonging, power, fun, freedom, and survival.

Glasser teaches that our basic needs are created from a set of genetic instructions that we must satisfy continuously. Four of the basic needs (love, power, fun, and freedom) are controlled by the cerebral cortex; Glasser calls them psychological needs because they are at a conscious level. The last basic need, survival, is controlled by a cluster at the top of the brainstem. The survival need operates at an unconscious level (Glasser, 1984).

The following is an adaptation of Glasser's (1984) five basic needs:

1. *Love/Belonging.* The majority of human beings are driven to be a part of society, to receive and give love, and to belong. Within societal, familial, or occupational spheres, it is an asset to ask clients how they are fulfilling their needs in these settings.

2. *Power (Achievement, Self-Worth, Recognition).* Our need for power is often expressed through competition; it can also be expressed in the achievement of something that gives us a sense of self-worth.

3. *Fun.* Glasser states that it is through laughing and having fun that children learn best; this need for fun continues throughout adult life.

4. *Freedom.* A need for freedom provides us with the opportunity to make choices in our lives that will allow us to fulfill our needs. The need for freedom is so important that entire countries go to war over it.

5. *Survival.* This is basic need controlled by the old (primitive) brain, which carries out the task of keeping our body functioning and healthy (breathing, sweating, digesting). Sometimes the old brain cannot function by itself and needs help from the new brain (the old brain sends a message to our new brain that recognizes thirst).

Reality therapy, based on choice theory, asserts the following principles:

- Glasser (1984) states that everything we do in our lives, be it good or bad, effective or ineffective, pleasurable or painful, crazy or sane, sick or well, is a way to satisfy powerful forces within ourselves (basic needs).

- We all have the same human needs, but our wants are unique and highly individualized (Glasser, 1984). We are all mandated to try to fulfill our needs and wants. Glasser (1994) explains that we cannot be alive without being driven; we have built-in genetic instructions, with a group of basic needs that we must satisfy throughout life.

- Wubbolding (1988) explains when we receive what we want, we are pleased; when there is a broken bridge between what we want and what we perceive we are getting, we begin to behave to rebuild the bridge. Wubbolding teaches all behaviors are an attempt to meet our needs when we are not getting what we want. Glasser teaches that all behaviors are total behaviors, including thinking, doing, feeling, and physiology, and therefore should be expressed in verb forms (when someone feels depressed, that individual's actions, physiology, and thinking are also depressing). Therefore, Glasser teaches, *people are depressing themselves* rather than being depressed (Corey, 1996).

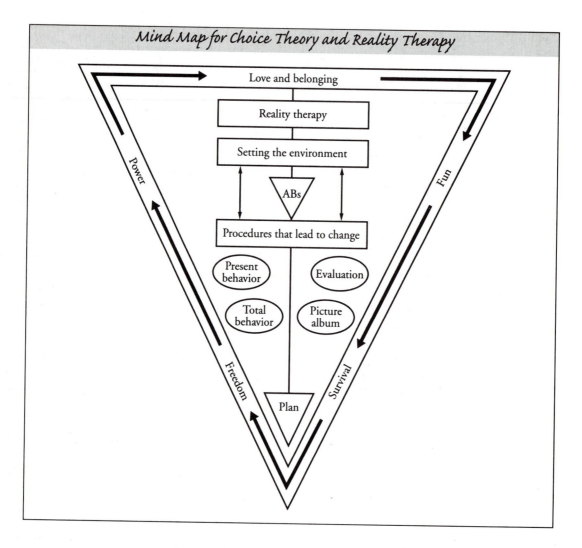

Mind Map for Choice Theory and Reality Therapy

- All human beings have individual desires. Glasser (1994) describes these desires as "quality pictures." A mother wants to hug her child. For the mother, the picture of hugging her child is a quality picture. Glasser (1982) explains that the picture album is located in our quality world. "Your picture album, in which you find love, worth, success, fun, and freedom, is the world you would like to live in, where somehow or other all your desires, even conflicting ones, are satisfied" (Glasser, 1984).

- Glasser (1965) points out one of reality therapy's concepts is *responsibility,* defined as the ability to fulfill one's own needs, but to do so in a way that does not deprive others of fulfilling their own needs (e.g., person giving love and receiving love). Therefore, individuals must take responsibility for their own actions.

- Wubbolding (1988) states reality therapy helps people take better control of their lives by accepting responsibility for their behavior. This therapy explains how we live our lives day to day. It is based on the viewpoint that the brain is a control system that looks for ways to mold the external world to satisfy our needs.

- Wubbolding (1988) states that one of the main principles of reality therapy is to be nonjudgmental as a counselor. Being able to look at a client's behavior without judging is crucial. The human services counselor should look at the client's behavior as the best the client was able to do at that particular time, given the information the client had. This does not mean the counselor has to agree with the client's behaviors (e.g., the client hit his wife). This client's behavioral system acted upon a need not being fulfilled—a difference between what he wanted and what he had. His behavior may have been the best he knew at that particular time.

Our genes, which are the building blocks of our lives, are nothing more than a series of instructions that we must follow if we are to survive and prosper. We become aware of many of these instructions as pictures in our heads, pictures that must be satisfied through the way we live our lives. Driven by our genes, we are captive to these pictures, but what we need to learn is that we are not captive to how we attempt to satisfy them. We almost always have choices, and the better the choices, the more we will be in control of our lives (Glasser, 1984).

TECHNIQUES

1. Setting the environment.

Wubbolding (1988) describes setting the environment as a very important component of reality therapy. It is meant to be a continuous process throughout the counseling relationship, forming a bridge to procedures that lead to change. The main goal of setting the environment is to enable the client to become comfortable with the counselor, so that an effective counseling process can occur. Wubbolding (1986) states that the client "has to know that you care before they can care what you know."

In reality therapy, Wubbolding (1988) notes that there are several guidelines to follow in establishing a safe environment for the client. They are described as the *always be (AB)* guidelines of counseling. One of the biggest advantages of reality therapy is that it offers the client choice theory. It is important to remember that reality therapy is a process and is flexible. To fully appreciate its elegance, you need to have a full understanding of the theory. The following sections provide a brief template of the main themes, so that you can start to use some of Glasser's work in your counseling orientation. Examples:

- Always be courteous
- Always be determined
- Always be enthusiastic
- Always be firm
- Always be genuine

2. Procedures that lead to change.

Exploring Wants and Needs. Wubbolding (1988) recommends exploration of the client's picture album (internal pictures of real wants), which provides an explanation that may assist the counselor in focusing on what the client truly wants. Other considerations include what family, friends, employer, counselors, and children may want from the client. Exploring the picture album (wants) is a continuous process according to Glasser (1994), because the pictures in the client's head continue to change daily. The more the counselor explores these wants, the clearer the picture becomes. This can be done by asking questions such as "What do you really want?" Example queries for exploring the picture album (Wubbolding, 1988) include:

- What do you want?
- What do you really want?
- If you had what you wanted, what would it look like?

Present Behavior. In using reality therapy with clients, the counselor focuses on present behaviors; the past is not ignored as long as it is associated with the present. Wubbolding (1988), discussing this aspect of Glasser, says "Everything we do today is in some way related to everything that has happened to us since birth." Reality therapy teaches we can only correct for today and plan for a better tomorrow. Glasser (1984) believes that in exploring present behaviors the counselor focuses on what the client is doing in the present tense, now, and should be looking at specific, unique, and precise actions. Examples of present behavior (Wubbolding, 1988):

- What are you doing?
- Tell me about your day, what did you do?
- When you were depressing yesterday, what were you doing?

Total Behavior. Total behavior is made up of four components: acting (doing), feeling, thinking, and physiology. Wubbolding (1991) provides an explanation of total behavior by using Glasser's analogy of the behavioral choice car (Figure 15–1). This car consists of four wheels , each wheel representing a component of behavior. All behavior is inseparable and total. Glasser (1984) postulates that whatever you do directly affects all four components of this synthesis. The choice car is a front-wheel drive. The front wheels of the car (thinking and acting) are the two components of behavior over which we all have some control. Glasser teaches that we have total control over our actions all of the time and control over our thinking some of the time. As with a front-wheel drive car, whenever the front components of behavior move, the other components of behavior have no choice but to follow. The client has indirect control over the rear components (feeling and physiology) by choosing an acting behavior For example, the client sitting in a room depressing may choose to get up and do an activity that individual really enjoys; as the client starts to engage fully in the activity, there is no choice except to move in a direction that leads away from depressing. Example for exploring total behavior:

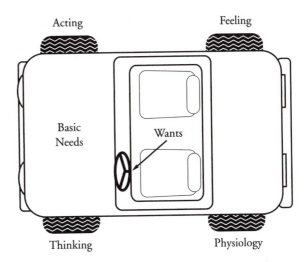

Figure 15–1.

Choice car (control car) (adapted from Glasser, 1984)

Acting

Feeling

Basic Needs

Wants

Thinking

Physiology

- "When you are with Johnny, what are you doing? What are you thinking about? What are you feeling? How does your body feel?" (*Note:* Ask questions one at a time, gathering information from the client's response before moving on. This process assists the client in discovering a greater degree of control over well-being than was originally thought).

Evaluation.   Evaluation questions help show clients they have control over their behaviors, since many clients believe they are controlled by their feelings. The counselor should never make value judgments for clients. Each client should decide individually what is important.

Wubbolding (1988) suggests it is important to focus on the client's specific behavior to ensure that what the client is presently doing is effectively fulfilling wants and needs. The counselor, by asking skillful evaluation questions, assists the client in learning if present behavior is, in fact, helping achieve desired goals. It is important to ascertain if the client's wants are realistic and attainable. It is also important to discover how committed that person is to the process of change. Examples of evaluation questions (Wubbolding 1988):

- If you continue to do what you are doing will you ever get what you want?
- Is what you're doing against the rules?
- Is what your doing helping you get what you want?

3. *Plan.*

Wubbolding (1988) contends the plan is a very important part of reality therapy, as well as the evaluation which takes place throughout the counseling process. Wubbolding (1988) recommends the following attributes should be taken into consideration when *formulating a plan with a client:*

- *Need fulfilling.* It is helpful and effective for the client to have a plan that is need fulfilling. If, for example, a client has a need for fun, an effective plan for this client will include fun.

- *Simple.* A simple plan, depending on the individual, may be lengthy or short. A short plan is useful for the client who is depressed. The proverb "A Journey of a thousand miles begins with one step" applies here.

- *Realistic and attainable.* The counselor helps clients keep the plan within their needs and capabilities.

- *Active.* A plan must be a "do-plan," a positive plan of action. For example, if a client states a desire to lose weight, a "do-plan" would include exercise.

- *Specific.* The plan should be specific, concrete, and exact. Where? When? What? With whom? How often? All these questions must be answered.

- *Repetitive.* The client should try to achieve something every day. This will help the client bridge the gap and move closer to attainment of desired outcomes.

- *Immediate.* A plan should be applied as soon as possible, preferably within the first 48 hours following the session. For example, the counselor might ask the client "What will you do tonight to change your life?"

- *Process-centered.* A process-centered plan focuses on the doer's action rather than on the outcome (e.g., process-centered: to go out Friday night; outcome plan: to meet people).

4. *Techniques that may be used in conjunction with reality therapy* (Please note: The following techniques are not a part of reality therapy; they are only tools that can be used in conjunction with it).

Paradox. A paradox is "a statement that is seemingly contradictory or opposed to common sense and yet is perhaps true" (*Webster's Ninth New Collegiate Dictionary*, 1989). Paradoxical techniques simply require viewing things in a different way, shedding a different light on problems. Wubbolding (1988, p. 69) describes the following example:

CLIENT: Yeah, I know. I'm bringing on the poor grades myself!

COUNSELOR: No, maybe the teachers are out to get you! (paradoxical statement)

The following guidelines direct the counselor when *not* to use paradoxical techniques:

- With sociopathic clients, because, when given a task, they do not process it.
- With paranoid clients; they may become suspicious.
- With clients who exhibit destructive behaviors.
- During acute crisis.
- Caution is advised when using paradox in families who exhibit chaos and confusion, so they do not push responsibility onto others (Wubbolding, 1988).

Reframing (relabelling, redefining). Wubbolding (1988) states this is a technique that can enable the client to look differently at a situation or topic. (*Note:* As with paradox, the counselor must be aware of proper use). For example, you are working with a client who is unemployed and you want to change the problem from serious to humorous:

> COUNSELOR: For today, instead of looking at yourself as unemployed, how would it be if you considered yourself to be on a one-day paid vacation?

Prescription. Wubbolding (1988) defines this technique as asking the client to choose the symptom, while the counselor defines the parameters. (*Note:* Counselors should not prescribe behavior such as anger.) Example:

> COUNSELOR: Try to be sad tomorrow from 8:30 to 9:30 only.

Listen for metaphors. A metaphor is "a figure of speech in which a word or phrase literally denoting one kind of object or idea is used in place of another to suggest a likeness or analogy between them" (*Webster's Ninth New Collegiate Dictionary* 1989). Wubbolding 1988 presents three purposes for using metaphors: (1) they give us greater understanding of what is already known, (2) they give us greater insight into the unknown, and (3) they help us express "that which has aesthetic and emotional relationship." For example:

- I feel like a doormat (low self-esteem).
- He acts like a bull in a china shop (aggressive).

Acknowledgments

The author is indebted to William Glasser and Robert Wubbolding for much of the material included in this chapter. The reader is referred to Glasser's 1998 *Choice Theory* and Wubbolding's 1991 *Understanding Reality Therapy,* both from Harper-Collins, in which their ideas are more completely delineated.

Summary

Reality therapy was created by William Glasser (1965) to provide a therapeutic model that will assist the counselor in moving the client forward. This is done by confronting the client's present behavior. As clients confront what they are doing, they move into a position to evaluate present behavior. Reality therapy encourages the client to take responsibility for life and learn to gain more effective control. One common theme is that we are internally motivated (internal locus of control). Choice theory provides ideas that allow reality therapy to work for the benefit of both the client and the counselor.

References

Corey, G. (1995). *Theory and practice of group counseling* (4th ed.). Pacific Grove, CA: Brooks/Cole.

Corey, G. (1996). *Theory and practice of counseling and psychotherapy* (5th ed.). Pacific Grove, CA: Brooks/Cole.

Glasser, N. (1980). *What are you doing?* New York: Harper & Row.

Glasser, W. (1965). *Reality therapy: A new approach to psychiatry.* New York: Harper & Row.

Glasser, W. (1984). *Control theory.* New York: Harper & Row.

Glasser, W. (1986a). *Basic concepts of reality therapy* (chart). Los Angeles: Institute for Reality Therapy.

Glasser, W. (1986b). *The control theory: Reality therapy workbook.* Canoga Park, CA: Institute For Reality Therapy.

Glasser, W. (1994). *The control theory manager.* New York: Harper & Row.

Glasser, W. (1995). *Staying together.* New York: Harper & Row.

Glasser, W. (1998). *Choice theory.* New York: HarperCollins.

Mowrer, O. H. (1947). On the dual nature of learning: A reinterpretation of conditioning and problem solving. *Harvard Educational Review 17:*102–148.

Webster's Ninth New Collegiate Dictionary. (1989). Markham, ON: Thomas Allen and Son.

Wubbolding, R. (1986). *Reality therapy training.* Cincinnati: Center For Reality Therapy.

Wubbolding, R. (1988). *Using reality therapy.* New York: Harper & Row.

Wubbolding, R. (1991). *Understanding reality therapy.* New York: HarperCollins.

Chapter 16

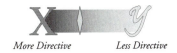

More Directive   Less Directive

Cognitive Therapy: Aaron Beck

Cognitive therapy is an active, directive, time-limited, structured approach . . . based on an underlying theoretical rationale that an individual's affect and behavior are largely determined by the way in which he structures the work (in his mind).

Aaron Beck, 1979

HISTORY

Aaron Beck, who was born in Rhode Island in 1925, is currently University Professor Emeritus in Psychiatry at the University of Pennsylvania. Beck is known worldwide as the father of cognitive therapy, which has developed a reputation as a short-term, cost-efficient form of psychotherapy. In addition, Beck is meticulous in measuring the effectiveness of his therapy through detailed research.

Cognitive therapy's theoretical underpinnings are derived from three main sources: (1) the phenomenological approach to psychotherapy, (2) structural and depth psychology, and (3) cognitive psychology. The development of cognitive therapy can be traced to Adler (see Chapter 11) and Piaget. Contemporaries in the psychotherapy field, such as Kelly, Ellis (see Chapter 17), Lazarus (see Chapter 18), and Mahoney were, along with Beck, beginning to give major recognition to *thought processes.*

Beck has published more than two hundred eighty titles. A study completed in *Canadian Psychologist* showed Beck to be the third most influential psychotherapist (second among those still alive—Albert Ellis being number one).

In the early 1960s, Beck was working as a psychoanalytical researcher with people who suffered from depression. Though originally intending to validate Freud's theory of depression, Beck's work evolved as a result of systematic clinical observations and experimental testing.

For more information about cognitive therapy, contact the Center for Integrative Psychotherapy, 1251 South Cedar Crest Blvd., Suite 211-D, Allentown, PA 18103.

MAIN IDEAS

Though much of Beck's research centered on serious clinical disorders, the basic principles and methods of his cognitive approach will be of use to counselors in a wide range of counseling situations. Especially with clients who have temporary and moderate disturbances, depressions, and anxieties, a focus on thoughts and

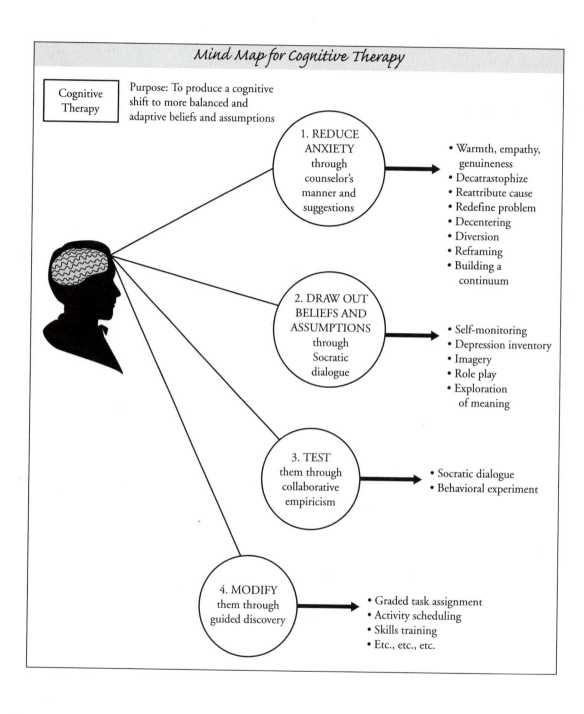

Mind Map for Cognitive Therapy

Cognitive Therapy

Purpose: To produce a cognitive shift to more balanced and adaptive beliefs and assumptions

1. REDUCE ANXIETY through counselor's manner and suggestions

- Warmth, empathy, genuineness
- Decatrastophize
- Reattribute cause
- Redefine problem
- Decentering
- Diversion
- Reframing
- Building a continuum

2. DRAW OUT BELIEFS AND ASSUMPTIONS through Socratic dialogue

- Self-monitoring
- Depression inventory
- Imagery
- Role play
- Exploration of meaning

3. TEST them through collaborative empiricism

- Socratic dialogue
- Behavioral experiment

4. MODIFY them through guided discovery

- Graded task assignment
- Activity scheduling
- Skills training
- Etc., etc., etc.

assumptions seems to make sense to people. Cognitive therapy in the professional context is largely a user-friendly approach, both for the counselor and the client.

In 1974, Beck and colleagues described how his work led to development of three specific concepts to explain depression.

1. *Cognitive triad.* Three elements represent the cognitive patterns of clients' outlook on life: how they view themselves, the future, and their world experiences (Figure 16–1).

2. *Increase in idiosyncratic schema.* A schema is a pattern of thinking, a cognitive structure, consisting of the individual's fundamental beliefs and assumptions (Beck and Weishaar, 1989). When appropriate and stable schema are upset by the intrusion of overly active schema unique to the individual, it can lead to a loss of voluntary control of thoughts and loss of ability to choose more appropriate schema (Beck et al., 1974).

3. *Faulty information processing.* Beck (1967) explains that clients' less effective thinking supports their belief that they are only capable of having negative things happen in their lives, even though there may be contradictory evidence readily available.

Cognitive therapy is based on the concept that the way we process information is important for human survival. Beck and Weishaar (1989) describe the ways our methods of processing information leads to various psychopathological concerns. These disturbances are caused by what Beck calls *systematic bias* when processing information. Like others who are oriented toward the cognitive realm, Beck believes his clients can consciously choose reason, and that counseling should be aimed at the underlying assumptions of the client himself. Beck notes the following areas in which his cognitive therapy differs from some other forms of psychotherapy (Beck and Weishaar, 1989; Beck, 1979).

First, the cognitive counselor interacts actively and purposefully with the client, structuring the session according to a particular plan. The goal of this directiveness is not to take over the client's freedom but, on the contrary, to stimulate the client

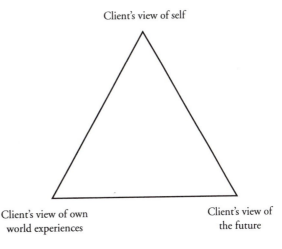

Figure 16–1.
Beck's cognitive triad (adapted from Beck and Weishaar, 1989)

Client's view of self

Client's view of own
world experiences

Client's view of
the future

to become actively engaged in the therapeutic process. Second, cognitive therapy focuses on "here-and-now" concerns, paying little attention to childhood problems, except to enhance the discussion of the present. Third, the strategy of cognitive therapy differs from some other cognitive orientations in its emphasis on the empirical investigation of a client's assumptions and automatic thoughts (those that come virtually without thinking). The counselor suggests ways that clients can test the validity of their thinking. Cognitive therapy is summed up in the phrase *collaborative empiricism,* since the key idea is the joint testing of a client's thoughts and assumptions by both the counselor and the client with the goal of trying to modify dysfunctional interpretations (Beck et al., 1974). The phrase also suggests the generally nonconfrontational manner in which the sessions are conducted.

In contrast to rational-emotive behavior therapy (Chapter 17), cognitive therapy views the client's problem not as philosophical (irrational thinking) but as functional (nonadaptive thinking) (Beck and Weishaar, 1989). Beck and colleagues (1974) explain that the overall purpose of cognitive therapy is to produce a *cognitive shift,* whereby clients move their thinking away from a particular bias or focus that brings about problematic behaviors to something more flexible, balanced, and adaptive (see accompanying mind map).

Cognitive therapy has been used effectively with depressed individuals, and it is certainly tailored to that particular concern. However, the same strategies and techniques have been shown to be useful in group work and when dealing with other concerns such as suicidal ideation, mania, anxiety or panic disorders, phobia, paranoia, hysteria, obsession, compulsion, anorexia nervosa, and hypochondria (Beck and Weishaar, 1989).

TECHNIQUES

As shown in the mind map, the techniques of cognitive therapy have been organized under four headings: (a) reducing anxiety, (b) drawing out beliefs and assumptions, (c) testing them, and (d) modifying them (to demonstrate the cognitive shift).

Reducing Anxiety

Beck believes it is important to reduce any immediate anxiety that a client carries into therapy (Beck and Weishaar, 1989), so that positive collaboration and cooperation can be established for dealing with the main concern. The following tools are used to promote a calm state in the client.

1. *Warmth, empathy, genuineness* (Todd and Bohart, 1994). For a full definition, see Chapter 8. These particular techniques help to create an atmosphere that is conducive to establishing trust. Examples:

   - Nonjudgmental attitude to the client and the problem.

   - Listening posture of the counselor's body and eyes.

   - Willingness on the part of the counselor to reveal appropriate emotion and not hide behind a professional mask.

2. *Decatastrophize* (Beck and Weishaar, 1989). Use a "what if . . ." question to help clients see that the problem may have grown in their mind, that it may not be the catastrophe it is thought to be. Examples:

- What if your mother did marry again? Would she automatically break the bond you have with her now?
- What if you did fail the entrance exam? Would that mean you couldn't try again next year—or apply somewhere else?
- Suppose you did have to declare bankruptcy. Wouldn't you be one of thousands each year who do so and live to see themselves, with the wisdom of experience, in business again?

3. *Reattribute cause* (Beck et al., 1974). Suggest that the problem has causes other than the one the client may be (wrongly) focusing on. Example:
   - Would a really bad mother come here of her own accord to deal with losing her temper at the children? It seems more plausible that you just have not yet found a way to relax after the daily frustration of your workplace.
   - Did your children really move down South just to spite you, or is it possible that jobs, pay, and the weather are better there?
   - Sometimes a nightmare, rather than indicating a predictive gift, is caused by eating fatty foods or milk products near bedtime. You say you often have pizza at night?

4. *Redefine the problem* (Beck and Weishaar, 1989). Put the problem in words that show the client what can be done about it. This will empower the client and diminish the image of doom with which the problem is associated. Examples:
   - You say you've tried everything to stop heavy perspiration when you're in front of an audience. Have you tried sweating on purpose—really trying to ooze? It does wonders to keep some people dry (Victor Frankl's paradoxical intention).
   - Next time you think any of the participants in the program are going to harm you, it would be a simple thing to go straight to a staff member.

5. *Decentering* (Beck and Weishaar, 1989). This is any technique used to persuade anxious clients that they are not the focus of everyone's attention. Examples:
   - You are very interested in what others think about you and how they look for ways to discredit you, but don't most of them lead busy lives with many other things on their minds?
   - I'm looking around at everyone here in the cafeteria and I only see three people looking at you . . . and one of them is me. Would you like to check it yourself?
   - At your next group meeting, why not ask how many people recall the gaffe you made at the previous meeting, and time them on how long it takes them to remember.

6. *Diversion* (Beck and Weishaar, 1989). This is any form of distraction that will break an emotion or thought pattern for a sad or anxious client. Examples:
   - [to increasingly distraught person] Forgive me, but I could not help noticing

just now what an unusual ornament that is on your necklace. Does it have a special significance?

- Join me for a walk in the courtyard.

- Tell me about one bit of humor that you could see in your situation (Beck et al., 1979).

7. *Reframing* (Beck and Weishaar, 1989). When clients reveal an either/or, black/white view of an event, have them place it in a different light. Examples:

- CLIENT: Either I get that done or I'm a total failure. Then I would have to quit my job.

- COUNSELOR: Who would say you are a failure?

- CLIENT: Mostly myself, I suppose. I couldn't face it.

- COUNSELOR: Is it fair to say you would be firing yourself?

- CLIENT: Maybe you're right.

8. *Build a continuum* (Beck et al., 1974). Generate middle-ground options for a person who has either/or, black/white thinking. Examples:

- COUNSELOR: Could you get help to ensure that you complete this crucial task successfully?

- CLIENT: This is too important. It will cost me my job if it gets botched. I have to do it myself.

- COUNSELOR: You say it is so important to you. Wouldn't that mean you can benefit by getting help?

- CLIENT: Well, I like to work independently, but yeah, I guess I should get a consultant.

Drawing Out Beliefs and Assumptions

When some trust has been established and immediate anxieties have been reduced, the task of discovering the client's beliefs and assumptions begins. This is done largely through bringing forward the person's *automatic thoughts*. To accomplish this, the counselor draws on an array of techniques, many of which involve the use of questions.

1. Self-monitoring

(Meichenbaum, 1977). It is often necessary to discover what thoughts a client has just before, or during, a problem behavior. This will provide a clue to the behavior and to its treatment. The client can monitor these thoughts and record them for discussion during counseling sessions. Examples:

- Client A records on a note pad angry and despairing thoughts occurring just prior to eating binges.

- Client B records on a wrist counter how many times he calls himself "dummy," "idiot," and similar expressions throughout the day.

- Client C records the number of times during the daytime she takes short naps on the couch or bed, and what thoughts she was having when she felt the need to lie down.

2. Depression inventory

(Beck et al., 1974). This is an important tool for assessing depressed clients. It is a 21-part questionnaire (reproduced in Beck, 1979), that is available for use by qualified therapists. The inventory questions are aimed at gauging the depth and extent of a person's depressive mood. (*Note:* Beck has also developed instruments to assess suicide, anxiety, and hopelessness.) Examples of inventory questions:

- Question 1.
    0. I do not feel sad.
    1. I feel sad.
    2. I am sad all the time and I can't snap out of it.
    3. I am so sad or unhappy that I can't stand it.
- Question 7.
    0. I don't feel disappointed in myself.
    1. I am disappointed in myself.
    2. I am disgusted with myself.
    3. I hate myself.
- Question 17.
    0. I don't get more tired than usual.
    1. I get tired more easily than I used to.
    2. I get tired from doing almost anything.
    3. I am too tired to do anything.

3. Imagery

(Beck et al., 1974). This involves the use of metaphors and pictures by both counselor and client. It may be used when the client has difficulty following ordinary verbal discourse, or in order for the counselor to get a clearer impression of the client's *automatic thoughts*. Examples:

COUNSELOR: How are you now?

CLIENT: Jumpy.

COUNSELOR: In what way?

CLIENT: Like a jackrabbit. You know, to avoid getting shot.

* * * * *

COUNSELOR: What were your thoughts after failing the test?

CLIENT: Dark.

COUNSELOR: You mean dark like twilight?

CLIENT: Midnight without a moon.

4. Role play

(Beck et al., 1974). Role play is acting the part either of someone else or oneself (usually as if in conversation with someone else). Role play is used for many purposes, in various orientations, but in this context is used to connect with *automatic thoughts*. Example:

> COUNSELOR: Let's pretend for a moment that I am your mother, and I have just commented unfavorably on your housekeeping. Respond to me the way you typically might to her.
>
> CLIENT: There it is again! Always criticizing me! Why don't you just leave me alone!
>
> COUNSELOR: Do I always criticize you?
>
> CLIENT: Often enough!
>
> COUNSELOR: That was genuine. Really, does she criticize you often?
>
> CLIENT: At least once every time she visits.
>
> COUNSELOR: How often does she visit?
>
> CLIENT: Well, she comes every Easter and Christmas . . .

5. Explore meanings

(Beck and Weishaar, 1989). This is an information gathering tool. It is used to check what the client's words actually mean, or it can be used to see what meaning an event has in the client's mind. Examples:

> COUNSELOR: When you say "Here we go again," are you saying that it is unfortunate but you'll go with it, or do you mean you feel you are being shafted?
>
> * * * * *
>
> COUNSELOR: People were not responding to you. Did you draw any conclusion from this?
>
> CLIENT: Well, sure! It's obvious I'm a loser.

Testing Beliefs and Assumptions

As automatic thoughts and their underlying assumptions begin to appear, the counselor enters into a partnership with the client to check on the accuracy of those statements. Each belief that the counselor thinks may be caused by unhelpful behaviors is treated as if it were a hypothesis that the client can test with the aid of the counselor. When the client sees that the assumption has no firm basis in reality, the groundwork is laid for a cognitive shift to a modified and more adaptive way of thinking.

1. Socratic dialogue.

This refers to a discussion generated by the counselor, through a thoughtful set of questions, that leads the client to arrive at logical personal conclusions (Beck and Weishaar, 1989). For example, a 26-year-old graduate student had a four-month history of recurrent depression (Beck et al., 1974):

CLIENT: I get depressed when things go wrong. Like when I fail a test.

COUNSELOR: How can failing a test make you depressed?

CLIENT: Well, if I fail I'll never get into law school.

COUNSELOR: So failing a test means a lot to you. But if failing a test could drive people into clinical depression, wouldn't you expect everyone who failed a test to be depressed? . . . . Does everyone who fails get depressed enough to require treatment?

CLIENT: No, but it depends on how important the test was to the person.

COUNSELOR: Right, and who decides the importance?

CLIENT: I do.

COUNSELOR: And so, what we have to examine is your way of viewing the test (or the way you think about the test) and how it affects your chances of getting into law school. Do you agree?

CLIENT: Right. (Beck et al., 1974)

2. Behavioral experiment

(Beck, 1979). In the context of testing beliefs, behavior experiments are designed to test a faulty hypothesis the client holds. In the following example, the assumption of the client is that people generally mock her and that they do so because she is foolish. The counselor questions the client, using Socratic dialogue, and ends with the formulation of a behavioral experiment. Example:

CLIENT: Part of me says I want to visit my childhood home on the coast . . . get in touch with roots, I guess. But everyone will laugh and talk me out of it. They think I'm so silly and sentimental . . . it is a kind of stupid idea, I guess.

COUNSELOR: You say everyone will laugh?

CLIENT: Yeah. They always do.

COUNSELOR: You told me. I didn't laugh, did I?

CLIENT: No, you didn't laugh because you're not supposed to as a counselor.

COUNSELOR: I didn't laugh because I thought your idea had merit. Suppose we agree to have you tell just two fairly neutral acquaintances tonight about your idea, just to see if they respond the way you think they will. Would you be willing to test your thinking this way and report back?

Modifying Beliefs and Assumptions

The encompassing technique of *guided discovery* is used initially in the behavioral experiment (see the previous section) and further, as the counselor serves as a guide, designing new experiences that lead the client to acquire new skills and outlooks (Beck and Weishaar, 1989). This part of cognitive counseling uses behavioral techniques almost exclusively. To acquaint yourself with possible techniques to complete the cognitive approach, see Chapter 18.

Acknowledgements

The author is indebted to Aaron Beck for much of the material included in this chapter. The reader is referred to Beck's 1993 *Cognitive Therapy and the Emotional Disorders* (New American Library) for a more complete delineation of his ideas. In addition, I recommend Marjorie E. Weishaar's 1993 *Key Figures in Counseling and Psychotherapy,* published by Sage Publications, for her chapter on Aaron Beck.

Summary

Cognitive therapy is a well-researched approach, developed especially for clients suffering from clinical depression. It also has a wide range of applications to other kinds of concerns. It is based on the belief how a person thinks largely determines how that person feels and behaves. It makes use of behavioral techniques in the later stages of the counseling process, but mainly employs dialogue between the counselor and the client to bring about joint discovery. The client's thoughts and assumptions are treated as hypotheses that can be tested, not for their "rationality" but for their flexibility and adaptiveness to life's difficulties. The techniques of cognitive therapy are particularly useful for reducing a person's level of anxiety, and the collaborative nature of the process further minimizes a client's discomfort.

Beck's *cognitive triad* and *depression inventory* are helpful tools to gauge the extent and depth of a client's sense of hopelessness and may alert the counselor to physically harmful tendencies in the client. The increasing recognition of the effectiveness of cognitive methods emphasizes the importance of the human service counselor's familiarity with these methods and their theoretical basis.

References

Beck, A. T. (1967). *Depression: Causes and treatment.* Philadelphia: University of Pennsylvania Press.

Beck, A. T. (1979). *Anxiety checklist.* Philadelphia: Center for Cognitive Therapy.

Beck, A. T. (1992). *Professional resume.* Philadelphia: University of Pennsylvania Press.

Beck, A. T. (1995). *Biography.* Philadelphia: University of Pennsylvania Press.

Beck, A. T., Rush A. J., Shaw, B. F., Emery, G. (1979). *Cognitive therapy of depression.* New York: Guildford Press.

Beck, A. T., Weishaar, M. E. (1989). Cognitive therapy. In R. J. Corsini, D. Wedding (eds.), *Current psychotherapies.* Itasca, IL: F. E. Peacock.

Beck, A. T., Weissman, A., Lester, D., Trexler, L. (1974). The measurement of pessimism: The hopelessness scale. *Journal of Consulting and Clinical Psychology 42*:861–65.

Emery, G., Hollon, S., Bedrosian, R. C. (Eds.). (1989). *New directions in cognitive therapy.* New York: Guildford Press.

Freeman, A. (Ed.). (1992). *Cognitive therapy with couples and groups.* New York: Plenum.

Meichenbaum, D. (1977). *Cognitive behavior modification: An integrative approach.* New York: Plenum Press.

Pace, N., Terry, M., Dixon, D. (1993). Changes in depressive self-schemata and depressive symptoms following cognitive therapy. *Journal of Counseling Psychology 40* (3):288–94.

Todd, J., Bohart, A. C. (1994). *Foundations of clinical and counseling psychology.* New York: Harper-Collins.

Chapter 17

More Directive   Less Directive

Rational-Emotive Behavior Therapy: Albert Ellis

The way people twist and convolute their thinking, emotions, and behaviors can be explained by ABC theory.

Ellis, 1958

HISTORY

Albert Ellis was born in 1913 in New York, and he grew up in the Bronx. Early in life he developed kidney disease and was hospitalized eight times between the ages of five and eight. He was subsequently forbidden outside activities. As a result, the introverted Ellis turned into a "stubborn and pronounced problem solver" (Hunt, 1993).

During his twenties, Ellis obtained degrees in accounting and business and, by the age of thirty-four, a doctorate in education, though he had a strong interest in writing books on sexuality. Following four years of psychoanalytic training, he established a full-time clinical practice in Manhattan by 1952.

Ellis rebelled against psychoanalysis almost immediately, saying he found it too slow, too passive, and unsuited to his own personality. He experimented during the next two years, and in 1955 developed rational-emotive therapy (RET). During this time Ellis had much opposition, but his ideas began to catch on in the early 1960s. By the late sixties, he had founded the Institute for RET in Manhattan.

In the Summer 1993 issue of the Institute for Rational-Emotive Therapy newsletter, Ellis announced that he was changing the name of his approach to rational-emotive behavior therapy (REBT) because the model had always stressed the reciprocal interactions among cognition, emotion, and behavior.

In 1982, a survey by *American Psychologist* found that 800 clinical and counseling psychologists regarded Ellis as the second most influential psychotherapist, behind only Carl Rogers (Hunt, 1993). Rational-emotive behavior therapy has pioneered a large number of thinking, feeling, and activity-oriented counseling programs. One such is Michael Bernard's "You Can Do It Program," an excellent program that guides parents and educators in teaching children to think. The program is useful for counselors as well.

For more information, write the Institute for REBT, 45 East 65th Street, New York, NY 10021-6593.

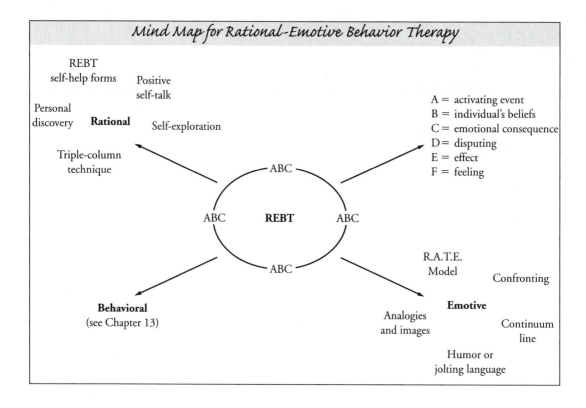

Mind Map for Rational-Emotive Behavior Therapy

REBT self-help forms

Positive self-talk

Personal discovery

Rational

Self-exploration

Triple-column technique

ABC

ABC   **REBT**   ABC

ABC

A = activating event
B = individual's beliefs
C = emotional consequence
D = disputing
E = effect
F = feeling

R.A.T.E. Model

Confronting

Behavioral
(see Chapter 13)

Analogies and images

Emotive

Continuum line

Humor or jolting language

MAIN IDEAS

The usefulness of rational-emotive behavior therapy to human services counselors is obvious. Its effectiveness is supported by research and its relevance has been tested on a wide range of personal concerns (e.g., unemployment issues, addictions, and vandalism) (Scissons, 1993). Human services counselors can apply REBT to any situation in which a client's thinking is self-defeating. Rational-emotive behavior therapy assists individuals in ridding their lives of inefficient, demanding thought patterns that lead to misery and disappointment.

The role of the REBT counselor is to help the client identify faulty self-talk that leads to negative self-fulfilling prophecies, and to help the client accept the self nonjudgmentally. As a result, the client learns to challenge rigid thinking and live a more balanced and productive life.

Rational-emotive behavior therapy, though focusing on thinking, does promote action. Rathus (1990) quotes Ellis as stating: "We need less misery and less blaming, but more action."

Cognitions are what we perceive, think, remember, interpret, imagine, and reason. They significantly influence our behavior. How people perceive certain events often determines their emotional reactions; the combination of these cognitions and reactions results in a behavioral response. As far back as the first century, the Stoic philosopher Epictetus wrote: "Men are disturbed not by things but by the view they take of them" (Meichenbaum, 1977). Rational-emotive behavior therapy is centered around the idea that all individuals are born with the potential to be totally rational as well as irrational (Ellis, 1989). This potential for irrationality stems from the individual's ability to interpret certain events unrealistically (hence, the term *irrational belief*). These irrational beliefs, Ellis suggests, are at the very core of an individual's shortcomings (Ellis and Grieger, 1986). In order for REBT to work, the client must assume responsibility for any disturbed thoughts, feelings, and behaviors. Clients must work on their ability to recognize these irrational beliefs when they are experiencing them and agree to make an effort to change the irrational belief that is causing the thought, feeling, or behavior (Corey, 1995a). Rational-emotive behavior therapy can be broken down into three distinct components using the letters A, B, and C (Figure 17–1); this is often referred to as the *ABC Model* (Rathus, 1990). *A* is the activating event, *C* is the emotional consequence and *B* is the individual's belief. Most people think that A causes C, but REBT suggests that while A contributes to C, it is B that is the most important determinant.

A Activating Event. This occurrence is what the individual becomes upset or disturbed about. It is only the starting point, not the real cause of the problem (Rathus, 1990).

B Belief. All problems stem from faulty thinking and in turn produce irrational beliefs or faulty assumptions. Assumptions or beliefs become faulty or irrational when they generalize, when they reinforce overall feelings of inadequacy, and or when they make sweeping statements about another person's character (Rathus, 1990).

Cavanagh (1990) explains that irrational beliefs seem to be mostly learned. The following examples describe five different ways this learning of irrational beliefs can take place.

- Direct experience: This learning involves events happening directly to the client (e.g., a young woman who has a series of bad dates, may generalize that males are typically insensitive and cruel).

- Vicarious experience: This learning evolves through association (e.g., the younger sister of the girl with the bad dates, though never having been on a date herself, picks up the thought from her sister that males are typically insensitive and cruel).

- Direct instruction: This learning is most like teaching (e.g., a girl is instructed by a favorite writer that males are typically insensitive and cruel).

Figure 17–1. Ellis' ABC theory of personality (adapted from Ellis, 1989)

Activating Event ⟶ Belief ⟶ Consequence

Disputing ⟶ Effect ⟶ New Feeling

- Symbolic logic: This learning uses representation (e.g., a child watches as anger destroys her parent's marriage, and from this experience the child concludes that all anger is bad and destructive and is to be avoided at all costs).

- Misinterpretation of cause and effect relationships (e.g., a boy may interpret that being held back a grade means he is stupid and slow, when in fact he was too young to be in that particular grade, and his parents and teachers felt it was best for the boy to be socializing with children his own age).

These errors in thinking, as Cavanagh (1990) states, form faulty beliefs; they produce certain consequences such as over-generalizations, all-or-nothing concepts, absolute statements, and poor time judgements. All lead to a behavioral response called a consequence.

C   Consequences. Consequences are realized in many aspects of a person's life (disturbed feelings and self-defeating behaviors are the most common). Consequences are the end result of an individual's faulty beliefs. Ellis suggests these behaviors are, at best, not constructive and, at worst, destructive to oneself and to others (Rathus, 1990).

D   Disputing. These ABCs of rational-emotive therapy can be joined by D, E, and F. The letter *D* stands for "disputing," which is the key therapeutic method. An REBT counselor intervenes in the destructive cycle by challenging the client's faulty beliefs and conclusions, and by helping the client to challenge self-talk and unrealistic thinking (e.g., "Where did you get the idea that you must get straight A's or you are a poor student?"). The course of action is to get the client to face the cognitive distortions that lead to a less productive behavioral pattern (Todd and Bohart, 1994).

E   Effect. The effect is the intended end-result of the application of REBT to a client. That effect produces a deep-seated cognitive change, namely, rational thinking (Todd and Bohart, 1994).

F   Feeling. The letter *F* refers to a new set of feelings; instead of feeling anxious or depressed, we feel appropriately in accord with a situation (Corey, 1996).

TECHNIQUES

Rational-emotive behavior therapy has one major goal: to decrease the client's self-defeating outlook and, consequently, to acquire a more realistic, tolerant philosophy of life (Ellis, 1989). The cognitive techniques that follow show how the human services counselor can use elements of REBT to achieve that goal. Here, the counselor uses a rationalization approach in assisting the client with the problem.

1. *Continuum* (Freeman, 1994). The use of a continuum can be very valuable in assessing the emotions and feelings of the client. The continuum can give a picture of how much the irrational thought, feeling, or behavior is affecting the person. Ask the client to rate a current feeling on a scale of one to a hundred (or whatever type of scale you would like to use) and then work toward decreasing the impact of that feeling on the client. Example:

Beginning Session

> COUNSELOR: On a scale of 1 to 10, how depressed would you say you feel after these incidents?
>
> CLIENT: I would say around 9.5.

A Few Sessions Later

> COUNSELOR: On a scale of 1 to 10, how depressed would you rate yourself after these incidents?
>
> CLIENT: I would say around 6.

2. *Teaching the ABC model.* The counselor teaches the client the basic model of REBT. During this process, clients learn how their own thinking causes their shortcomings (Ellis, 1973). Example:

> CLIENT: I got turned down for the loan (activating event). There must be something wrong with me (faulty belief). I must be worthless (consequences: negative self-talk).
>
> COUNSELOR: Where is it written that if someone gets turned down for a loan there must be something wrong with them? (disputing the belief)
>
> CLIENT: Nowhere I guess, it's just something I think about myself.
>
> COUNSELOR: [Teaches the clients the ABC model.]

3. *Personal discovery.* Challenges the client to do selected self-help exercises that help achieve deep-seated cognitive change. Through exploration, disturbance-creating ideas can be eliminated and minimized by a vigorous dispute with logical-empirical thinking (Cavanaugh, 1994). For example, the counselor assigns the homework task of keeping a daily record of every person who talks to the client until the next session. The client previously claimed that no one ever noticed them.

4. *Self-exploration.* With the help of the counselor, the client is asked to explore certain thought patterns that lead to self-defeating behaviors. This method gives the client a chance to examine personal irrational thinking (Ellis and Grieger, 1986). Example:

> CLIENT: I know that getting turned down for that loan was not what I wanted to hear, and it did make me feel really bad, but it doesn't mean that I'm worthless, or any less of a person. I felt a negative feeling that was based on a belief of being worthless. I can now change my old belief to a more positive and rational belief (e.g., bank loans do not evaluate a person's inner worth).

5. *REBT self-help form.* This form is used by the counselor to determine the faulty beliefs that influence clients, and to what extent they do so. The form is completed by the clients, which leads them, on their own, to recognize situations that are causing a problem. The form is used to record the event (A of the ABC Model) that contributed to the upset. Then the person would record the

consequence (C), either the feeling or behavior that the person would like to change. After the event and consequence have been identified, the form is used to identify any irrational beliefs the person holds about the event. In following the ABC Model, the belief is then disputed and replaced with a rational one (Corey, 1995a). These REBT self-help forms can be obtained from The Institute for Rational-Emotive Behavior Therapy (address in history section at the beginning of this chapter).

6. *Positive self-talk.* The counselor challenges the client to turn self-defeating self-talk into thought patterns that are more positive and productive (Ellis, 1973). Example:

CLIENT: I can't get anything done, I'm so disorganized, and I'm slow.

COUNSELOR: Let's try to change some of that negative self-talk. Try saying this: *I can get things done, I am organized, and I'm precise.*

CLIENT: I can get things done, I am quite organized, and I am very precise. Yeah, I like that.

7. *Triple column technique.* This technique utilizes a table with three columns and is aimed at detecting irrational thoughts and beliefs. Its goal is to substitute more rational, objective thoughts for the illogical or harsh self-criticism the person may currently create (Burns, 1988). For example, the emotive part of REBT deals with the client's feelings. Feelings can be used as a way to develop a relationship with the client or to point out inconsistencies between the way clients feel and the way they think and behave.

Automatic Thought	Cognitive Distortion	Rational Response
"I am dumb!"	"I should be able to do this every time without a problem."	"I am human, and I do not have to be perfect every time."

(*Note:* This technique is also used by cognitive therapists.)

8. *Humor.* A skilled counselor can use humor to show the client that certain beliefs or thoughts are ridiculous, absurd, faulty, or irrational. Humor may be achieved, for example, by using deliberately strong language to shift clients existentially so they loosen up and open up. Ellis often used humor, sarcasm, and strong language if the client seemed responsive to these, even to the point of composing comic verse (Arbuckle, 1967). (*Note:* Be careful not to offend the client; you must be sensitive to the client's possible negative interpretations.) Example:

CLIENT: I couldn't go into the bank wearing muddy work pants. Everyone would stare at me.

COUNSELOR: Next time you go into a bank, purposely wear muddy pants, then stop and count how many people are actually staring at your pants.

9. *Confronting.* A counselor uses this technique to highlight discrepancies in the client's thinking and actions. This method is an invitation for the client to self-evaluate internal or external behavior and self-defeating thought patterns (Ellis, 1989). Example:

COUNSELOR: You are telling me that you like school, but your principal is telling me that you have a high truancy record.

10. *Unconditional positive self-regard.* The counselor assists the client to have unconditional positive regard for self, independent of anyone else. The counselor helps the client see how behaviors have been unproductive. The emphasis is for the client to gain self-acceptance. Example:

CLIENT: I am a good person who is having a challenging period in my life. I have made a few mistakes, but who hasn't. I am OK, and I will be fine, no matter what.

COUNSELOR: Excellent! Self-acceptance is a big step to self-fulfillment.

11. *Forceful coping statements.* This is a technique Ellis developed to help the client change, not only thinking but also feelings. It takes commitment and the full effort of the client. Example:

CLIENT: [Writes down a statement she wants to feel good about.] At work, I'll never, never *need,* only prefer.

12. *Analogies and images.* Using these to illustrate problems allows the client to have a clearer visual picture or a different perspective of the presenting problem. Sometimes it is easier to see things when they are applied to something other than yourself (Vernon, 1995). Example:

- COUNSELOR: If you are driving a car and you get a flat tire, does that mean you throw the car away, or do you fix the tire?
- CLIENT: Fix the tire.
- COUNSELOR: Then if there is something that you can't do, do you give up on yourself and say that you can't do it, or do you fix it by working hard?

Having worked with the cognitive aspects of REBT, we now turn to the behavioral aspects. It is clear that thoughts, feelings, and beliefs affect behavior. When an irrational belief changes, a concomitant change in behavior will occur. In theory, this process also works in reverse. Sometimes changes in behavior can alter the thought patterns and beliefs of the client. Chapter 18 presented some techniques that can be used to modify behavior. Here is an example of a four-step model that can be used with REBT.

The *R.A.T.E. Model* (Vernon, 1995) was developed to put therapy into the context of the client's life. The *R* is for *relationship building,* which involves establishing a rapport with the client. This can be done by creating a comfortable environment and learning the personal interests of the client. The *A* is for *assessment.* There are two kinds of assessment. Assessing emotions and behaviors can be done by looking at all possible emotional reactions to many different situations, assessing the consequences, and exploring the connection between emotions and behaviors. Assessing of cognitions can be done by distinguishing between facts and assumptions and exploring inferences. The *T* is for *treatment*. The goal of therapy is to reduce the intensity of negative emotions and modify self-defeating behaviors. Many different treatment methods can be used. The *E* is for *evaluation*. This is an overall evaluation of the treatment process and its effectiveness. It involves subjective, objective, and independent observations by both parties.

Summary

I have included rational-emotive behavior therapy in this text because it provides the professional with a user-friendly directive orientation. For example, Ellis's ABC theory of personality enables many clients to examine the origins of their feelings and frustrations and rethink their old paradigms.

Rational-emotive behavior therapy provides the human services counselor with another orientation that links thinking, feeling, and behavior. The counselor who understands this approach may adopt part or all of REBT into a personal counseling orientation.

As in cognitive therapy, the goal of REBT is for clients to understand that they are at cause for their own thinking and, with practice and a concerted effort, to be able to overcome life problems. Rational-emotive behavior therapy provides a fund of educational benefits for lifelong self-management.

References

Arbuckle, D. S. (1967). *Counseling and psychotherapy: An overview.* New York: McGraw-Hill.

Burns, D. (1988). *Feeling good: The new mood therapy.* New York: Signet.

Burns, D. (1990). *The feeling good handbook.* New York: Penguin.

Cavanaugh, M. (1990). *The counseling experience.* Prospect Heights, IL: Waveland Press.

Corey, G. (1995a). *Student manual for theory and practice of group counseling* (4th ed.). Pacific Grove, CA: Brooks/Cole.

Corey, G. (1995b). *Theory and practice of group counseling* (4th ed.). Pacific Grove, CA: Brooks/Cole.

Corey, G. (1996). *Theory and practice of counseling and psychotherap,* (5th ed.). Pacific Grove, CA: Brooks/Cole.

Ellis, A. (1973). *Humanistic psychotherapy: The rational-emotive approach.* New York: McGraw-Hill.

Ellis, A. (1980). Rational-emotive therapy and cognitive behavior therapy: Similarities and differences. *Cognitive Therapy and Research 4:*325–40.

Ellis, A. (1989). Rational-emotive therapy. In Corsini, R. J., Wedding, D. (eds.), *Current psychotherapies* (4th ed.). Itasca, IL: F. E. Peacock.

Ellis, A. (1996). *Better, deeper, and more enduring. Brief therapy: The rational-emotive behavior therapy approach.* New York: Brunner-Mazel.

Ellis, A., Bernard, M. E. (Eds.) (1985). *Clinical applications of rational-emotive therapy.* New York: Plenum.

Ellis, A., Grieger, R. (Eds.) (1986). *Handbook of rational-emotive therapy.* New York: Springer.

Freeman, A. (1994). *Depression: A cognitive therapy* [pamphlet]. *Assessment and Treatment of Psychological Disorders* [video series]. New York: Newbridge Communications.

Hunt, M. (1993). *The story of psychology.* New York: Doubleday.

Meichenbaum, D. (1977). *Cognitive-behavior modification.* New York: Plenum Press.

Rathus, S. (1990). *Psychology.* Orlando: Holt, Rinehart, & Winston.

Scissons, E. (1993). *Counseling for results: Principles and practices of helping.* Pacific Grove, CA: Brooks/Cole.

Todd, J., Bohart, A. (1994). *Foundations of clinical and counseling psychology.* New York: Harper-Collins.

Vernon, A. (1995). Rational-emotive behavior therapy with children and adolescents. Paper presented at the Rational-Emotive Therapy Workshop, Dartmouth, Nova Scotia, November.

Chapter 18

More Directive   Less Directive

Behavioral Therapy: An Overview

Behavioral therapy shares many commonalities with other psychological therapies, particularly those that tend to be briefer and more directive.

Wilson, 1989

HISTORY

Behavioral approaches to counseling were not formulated systematically until the late 1950s, though they are based on theoretical and experimental work done earlier by behavioral scientists (Rachlin, 1970). There are a variety of behavioral schools falling on a continuum from one pole, where the "mind" is considered to be of no scientific interest, to the opposite pole, where personal thought is the key to all behavior. A common mistake is to judge all behavioral approaches on the basis of one or two controversial methods.

B. F. Skinner represents one end of the continuum of behavioral approaches. Skinner's goal was not to understand the inner person but to discover how behavior is created by external causes (Hunt, 1993). At the other end of the continuum lies the theoretical work of Julian Rotter, which focuses on internal causes (Goldfried and Davison, 1976). The behavioral techniques described in this book grow out of Rotter's theory of internal locus of control. In fact, these techniques overlap with the cognitive approaches of Beck (Chapter 16) and Ellis (Chapter 17).

The common ground of all behaviorists, according to Goldfried and Davison (1976), is the belief that behavior follows specific laws much as nature in general does. These approaches in counseling are thus founded on considerable empirical research and documentation.

For more information, write the Association for Advancement of Behavior Therapy, 305 Seventh Ave., New York, NY 10001.

MAIN IDEAS

The human services counselor does not have to accept all the ramifications of behavioral philosophy in order to use the techniques successfully. Indeed, behavioral methods are employed by a number of other therapies and can be readily combined with other approaches. In fact, there are behavioral techniques for a diversity of behaviors and learning concerns, as well as for many special populations. Behavioral tools are used today by school teachers and assistants, business managers, coaches, caregivers for the disabled or elderly, and many others. For a more complete range of behavioral approaches and their researched benefits, see the references for related literature in your field.

Certain assumptions and beliefs characterize all behavioral approaches. Wilson (1989) lists these perspectives as follows:

- Many abnormal behaviors, rather than being illnesses or symptoms of illness, are better thought of as "problems of living."

- Most abnormal behavior is acquired and maintained in the same way as normal behavior, through learning.

- Assessment of behavior deals with present causes rather than origins in the past.

- Analysis of the problem involves dividing it into component parts, so that treatment procedures can be targeted at specific components.

- Treatment strategies are tailored to different problems in different individuals.

- Successful treatment of a psychological problem does not require an understanding of its origins.

- Behavior therapy involves a commitment to the testable, scientific approach.

TECHNIQUES

Because behavior therapy tailors its approach to the individual situation, it has numerous methods. Following are twenty-two commonly used behavioral and cognitive-behavioral techniques, in alphabetical order. These were chosen for their appropriateness to counseling in the human services field and characterized in terms of their level of directiveness.

1. *Activity scheduling* (*moderately directive*) (Wilson, 1989). Setting up a schedule of activities is a way of moving someone from a listless, inactive state to being more productive and alive. Though the activities themselves imply behavior, the act of scheduling is largely cognitive and is intended to have some immediate positive effect on the person's reasoning by itemizing choices and strategies. For example, with an indecisive person, schedule what is needed to reach a decision: list things that need more information, i.e., go to the appropriate agencies or persons; compile data gathered; make a chart (pros and cons); set a time frame within which a decision is to be made.

2. *Assertion training* (*fairly directive*). A collection of techniques is drawn upon to teach an alternative to aggressive or passive behaviors. Assertiveness is the ability to express one's needs and thoughts confidently, without hiding or muting on one hand, or forcing and badgering on the other (Meichenbaum, 1977). Assertion training focuses on self-talk, a verbal or mental technique by which one

Mind Map for Behavioral Therapy

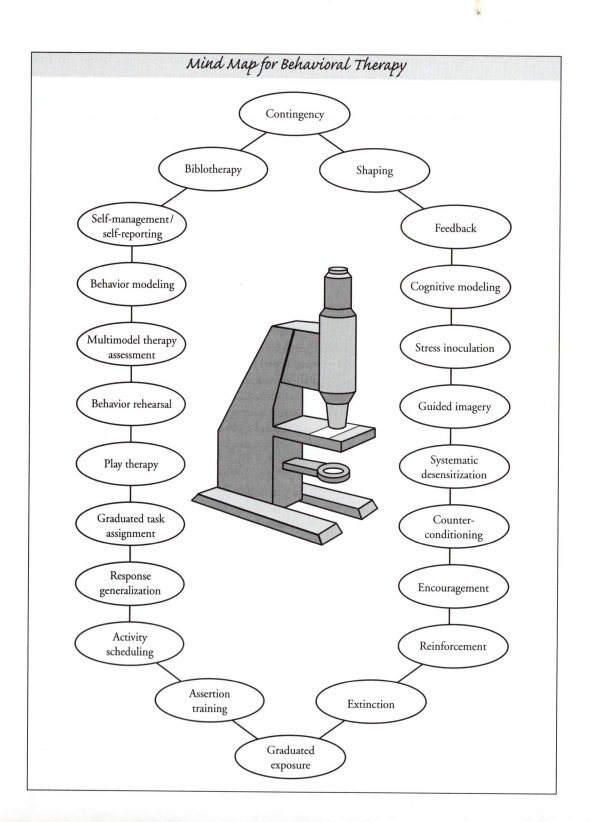

can modify thoughts about self or others. (Assertion training includes behavioral rehearsal, exposure, modeling, and reinforcement.)

Assertion training is useful to those who cannot express anger or frustration, those who have difficulty saying no, those who allow others to take advantage of them, those who have difficulty expressing affection, and those who feel that they do not have the right to express their thoughts and feelings. The goals of assertion training are:

- To empower people to make the choice to behave assertively in certain situations
- To teach people to express themselves in a way that reflects sensitivity to others
- To focus on client's self-statements, self-defeating beliefs, and faulty thinking
- To provide realistic opportunities for people to face and challenge difficulties in a safe environment.

Examples:

- Positive self-talk

CLIENT: My boss has a heartbeat, I don't need to be afraid; I can stand my ground.

- Use "I" statements

CLIENT: "When you . . . I feel . . . because . . . ," (e.g., When you talk when I am teaching, I feel frustrated because I think your talking is disturbing the class).

3. *Behavior modeling* or *vicarious learning (fairly to moderately directive)* (Schwitzgebel and Kolb, 1974). The client observes the counselor or someone else behaving or performing a task in order to learn from it or imitate it. When modeling is done in a social setting (e.g., the family) it is called "social learning." For example, in a demonstration of responding calmly to an agitated individual, clients observe the counselor remaining quiet and relaxed as another client yells at the counselor.

4. *Behavior rehearsal (fairly to moderately directive)*. New behaviors are practiced in a context that simulates the real world; it is a gradual shaping process. This helps the client to perform the desired behavior outside the therapeutic environment. For example, a client is encouraged to practice what she wants to say to her boss while standing in front of the mirror (Corey, 1996).

5. *Bibliotherapy (fairly directive)* (Pardeck, 1993). Bibliotherapy is defined as the use of literature to help clients cope with changes or problems in their lives. The counselor must be sensitive to the client's needs and use skill and insight in the selection process; the presentation of literature is based on a carefully planned strategy that ensures maximum benefits; during the follow-up stage the client is encouraged to share what was gained. The goals of bibliotherapy are:

- To provide information about the problem
- To provide insight into problems

- To stimulate discussion
- To convey new values and attitudes
- To normalize the problem (create an awareness that others have dealt with similar problems)
- To provide possible solutions

For example, when working with clients, literature or poetry offered can be useful for treating problems in the following areas: blended family, separation and divorce, child abuse, foster care, adoption, addictions. Reading a story based on such a theme can help a client open up and discuss feelings.

6. *Cognitive modeling (fairly directive)* (Meichenbaum, 1977). The counselor performs a task while talking aloud to self (self-talk or self-encouraging), to demonstrate how the client may talk the self through a task successfully. Cognitive modeling is a main feature of what is called self-instructional training. For example, encourage self-talk in a person with low self-esteem who needs to confront an intimidating coworker, to help combat fearful thoughts and promote confidence. The counselor provides an sample dialogue modeling one way the client might address the situation. The counselor repeats it until the client has grasped the concept.

7. *Contingency contracts (directive)* (Rose and Edlesson, 1987). A contingency contract specifies behavior to be changed, performed, or discontinued; rewards associated with achievement of stated goals; conditions under which rewards are to be received; and a time frame (as possible). Contracts are used to reduce or eliminate undesired behaviors, often with children; some adults find contingency contracts patronizing. The client is fully informed and actively involved in deciding on behaviors and rewards. Principles of contingency contracting include:

- Clear descriptions of specific behaviors to be performed
- Specifications of immediate reinforcement
- Descriptions of how goals will be observed, measured, and recorded

For example, a counselor and client may contract that the client will not eat any chocolate for a week; if successful, the client can have one chocolate bar during the second week. The contract will be monitored through a daily journal and reviewed once a week in counseling.

8. *Counter-conditioning (fairly directive)* (Wilson, 1989). Also known as *reciprocal inhibition* (Wolpe, 1990), this technique is a way of lowering or eliminating anxiety by having the client practice or experience an opposite emotion. Examples:

- Taking a deep breath, stretching and relaxing shoulder and neck muscles to counter a fearful tightness in the body.
- Singing or whistling to counter anxiety.

9. *Encouragement (moderately directive)* (Schwitzgebel and Kolb, 1974). An indication of approval or encouragement by the counselor for the purpose of helping the client deal with a problem or practice a preferred behavior. Examples:

- Verbal encouragement: "I can see that you'll soon get it, you're doing fine."
- Active encouragement: "Because you were able to speak to your boss, I encourage you to do something you really enjoy today."

10. *Extinction (moderately directive)* (Rachlin, 1970). The discouragement of an unwanted behavior and its eventual elimination by removing something that seems to stimulate or reward that behavior. For example, a parent stops fussing over a child when the child throws a tantrum for not getting her own way.

11. *Feedback (moderately directive)* (Rose and Edlesson, 1987). Feedback is a useful part of learning new behaviors if it is constructive, specific, and positive. Feedback should consist of both praise and encouragement, and include specific suggestions for modifying errors. Example:

- Positive feedback is given first.
- When criticism is necessary, focus on behaviors using "I" statements, and be specific in saying what could be done differently.
- Have the client express personal feelings about the feedback.

12. *Graduated exposure (also called* in vivo *training) (directive)* (Todd and Bohart, 1944).

This technique involves exposure to a real-life situation in a step-by-step procedure, in order to practice a skill or a preferred behavior in an anxiety-provoking environment. Examples:

- A mentally challenged person is afraid to cross the street. The counselor has the client progressively practice safe crossing techniques indoors, in the yard, in the parking lot, on a quiet street, at a crosswalk on a busy street. Finally, the client does it alone with the counselor watching.
- A nervous valedictorian practices a short speech in front of a small group, followed by a full speech, and then a speech on stage before a small group.

13. *Graduated task assignment (directive)* (Wilson, 1989). This is an assigned task, often as homework, which begins simply and proceeds in steps that become more difficult and complex. For example, here is an assignment for meeting people:

Step 1. Say "Hello" to a new person.

Step 2. Say "Hello" to two people consecutively.

Step 3. Say "Hello" to several people in one location.

Step 4. Say "Hello" to someone and make a comment about the weather.

Step 5. Say "Hello" and introduce yourself.

Step 6. Introduce yourself, and ask if you can be of assistance.

Note: Steps are done one at a time. Do not move to the next step until the client is able to succeed at the present step.

14. *Guided imagery (directive)* (Murdock, 1987). Imagery is used in guiding the client into experiencing relaxation or envisioning options for a hopeful future.

It involves creating pictures in the mind for the purpose of lowering anxiety, discovering alternatives in thinking, or increasing body awareness and self-esteem. Examples:

Part of an exercise to enhance self-expression

COUNSELOR: You climb up into your own spaceship and prepare for takeoff . . . you gently lift above the clouds . . . up above the earth's atmosphere, way out into space . . . you choose a planet or star to investigate and head your spaceship toward it. After landing your spaceship, you learn as much as possible about how the inhabitants of that planet live and communicate. (The client later reports what he or she imagined.)

Part of an exercise on the meaning of friendship (for adolescents)

COUNSELOR: Imagine you are traveling through space and time with a friend of your own choosing . . . You have chosen this friend to accompany you for a particular reason. What is it about this person that you like? Notice how you interact with this person. What it is about this friendship that you value?"

15. *Multimodel therapy assessment (moderately directive)* (Lazarus, 1989). The counselor who is helping a client change behavior must first be aware in which component of life the client is experiencing difficulties. Lazarus's BASIC I.D., which follows, is a model to help clients assess this. It also fits the spirit of this text, because Lazarus believed that all people have unique needs and problems ad that a professional must have an eclectic orientation in order to help more people. Lazarus looked at people through seven different components:

Behavior—refers to typical behaviors; ways a person acts that can be observed and recorded.

Affect—refers to how the client feels.

Sensation— refers to the five senses (taste, smell, touch, sight, and hearing).

Imagery— refers to how the client sees the self, including dreams, fantasy, and memory.

Cognition—refers to self-talk, values, beliefs, and opinions.

Interpersonal relationships—refers to interactions with others.

Drug/Biology—refers to drug use, diet, exercise, and overall health.

Examples:

B   John, what would you like to stop doing? John, what type of behavior do you do that causes you problems?

A   John, on a scale of one to ten, how emotional are you? What makes you happy or sad?

S   John, do any smells, touches, tastes, sounds, or sights cause you problems? In regard to the senses, do you have any you really enjoy?

I   John, what type of imagination do you have? How do you see yourself?

C John, what is some positive and negative self-talk you have? What rules do you live by?

I John, how do you get along with your family? Are there any relationships you would like to improve?

D John, how is your health? John, what drugs are presently or previously used?

16. *Play therapy (directive)* (Wilson, 1989). The practice of relaxation or preferred behaviors through a game that brings internal rewards (e.g., fun activity); this is especially suitable for children and for people who are mentally challenged. Examples:

   • Copycat game: Any game which focuses on having people copy behavior (e.g., Simon Says). Behavior will be learned through modeling.

   • Talk to the clown (or teddy bear or puppet): The child expresses feelings to a friendly-looking stuffed figure. (Computerized versions may be used, enabling verbal interaction.)

17. *Reinforcement, punishment, and omission (directive)* (Goldfried and Davison, 1976). People sometimes confuse the term *negative reinforcement* with punishment. In behavioral psychology these terms have technical meanings that differ (Figure 18–1).

   Punishment (aversive conditioning) involves presenting an unpleasant stimulus to produce a desired outcome (usually to decrease undesirable behavior); this technique is seldom used by those whose focus is internal locus of control. Examples:

   • A medical therapy that involves the client's taking the drug Antibuse, which causes nausea and vomiting when alcohol is consumed, with the result that the client stops drinking (at least while on the drug).

   • A parent who yells at the children to stop them from horseplay.

   Reinforcement involves the use of a pleasant stimulus ("positive reinforcement") or the removal of an unpleasant stimulus ("negative reinforcement"). Examples:

   • COACH: You've earned it today! My treat at the coffee shop. (Positive reinforcement because a pleasant stimulus or reward is presented.)

Figure 18–1.

	Stimulus Present	Stimulus Absent
Pleasant Stimulus	Positive Reinforcement	Omission (Extinction)
Unpleasant Stimulus	Punishment	Negative Reinforcement

- COACH: You've earned the right to quit practice early without the usual laps around the track. (Negative reinforcement because an unpleasant stimulus is removed.)

Omission occurs when a pleasant stimulus that is usually present is absent after a particular behavior. This is similar to extinction. Examples:

- A child is sent to his room after doing something bad (the delivery of a pleasant event is withdrawn, e.g., playing with his friends, watching television).

- Someone's license is suspended for drunken driving (withdrawal of the reinforcement or privilege of driving).

18. *Response generalization (directive)* (Rachlin, 1970). In this technique, the client is encouraged to demonstrate a newly learned behavior in a setting somewhat different from the one in which it was learned (yet with a few similarities), in order to make the behavior applicable to a variety of situations. Examples:

- A client has significantly decreased stuttering in a controlled environment; now he is asked to go across the road to the coffee shop and order a snack.

- A mentally challenged client has learned how to use a telephone in the workshop office and is now taken to a shopping mall to try a pay phone.

19. *Self-management and self-reinforcement (fairly directive)* (Watson and Tharp, 1993). Self-management can be applied to problems such as anxiety, depression, and pain; controlling smoking, drinking, and drugs; obesity and overeating; and learning study and time-management skills. The basic idea is that change can be brought about by teaching people to use coping skills in problematic situations. Generalization and maintenance of outcomes are strengthened by encouraging clients to accept responsibility for implementing these strategies in daily life. Examples (adapted from Watson and Tharp, 1993):

- Specify desired changes at the outset. Establish goals, one at a time, that are measurable, attainable, positive, and significant *for the client*. Consistency is essential.

- Translate goals into target behaviors. "What specific behaviors do I want to increase or decrease?" (Ensure that the client has the skills to carry out behaviors. Do not assume she knows what being happy looks like or how to act happy.)

- Practice self-monitoring. This means deliberately and systematically observing one's own behavior. It leads to awareness focused on concrete, observable behaviors. A behavioral diary records an event with its relevant antecedent cues and consequences, e.g., entries of what you ate, events prior to eating, frequency of eating, type of food eaten, etc. (*Note:* We believe a daily journal of behavior is very useful. The exercise of writing about activities and daydreams provides the client with daily evidence and research that can later be reviewed.)

- Form a plan. After evaluating desired behavioral changes, the client devises an action program to actually bring about change; this helps to replace an undesirable behavior with a desirable one. (See Chapter 15 for more information on planning.)

- Practice self-reinforcement. It is important to choose appropriate, personally motivating self- rewards. This is a temporary strategy to be used until clients can achieve new behaviors in everyday life.

- Do self-contracting. Determine in advance what external and internal consequences will follow from carrying out the desired or undesired behavior.

- Evaluate the plan for change. Determine the degree to which the goals were reached. Readjust and revise the plan of action on an ongoing basis as clients learn different ways to achieve their goals.

20. *Shaping (directive)* (Skinner, 1971). There are two types of shaping: forward (working toward a performance with reinforcement of closer approximations for the desired behavior) and backward (starting from a target performance and working back to your initial step). The purpose of shaping is to reach a desired behavior by using reinforcement of successive approximations to the required response, with gradual cessation of reinforcement of earlier responses (a form of operant conditioning). In the beginning processes of shaping, reinforcement is essential; it must continually reinforce behavior "each and every time it occurs." Praise is a simple but powerful reinforcer. Examples include the following.

- Teaching a child how to swim (forward shaping) using reinforcement:
  (1) The child enters the water.
  (2) The child gets his face wet.
  (3) The child is able to open his eyes under the water.
  (4) The child establishes breath control underwater.
  (5) The child front floats, assisted.
  (6) The child front floats, unassisted.

- Teaching a child to put on a pair of pants (backward shaping) using reinforcement:
  (1) Put the pants on the child up to the knees with the child having to pull his/her pants up.
  (2) Put the child's right leg in pants with the child having to put his/her left leg in and pull up the pants.
  (3) Put the child's left leg in pants with the child having to put in his/her right leg, and pull up the pants.
  (4) Pass the pants to the child to be put on with no assistance.
  (5) Leave the pants on the bed to be put on by the child with no assistance.

- The client is a three-year old boy with schizophrenia, lacking normal verbal and social skills. After having cataract surgery, he refused to wear the glasses that were essential to development of normal vision. The child was trained to expect candy or fruit at a clicking sound that became a conditional reinforcer. He was reinforced with the clicking sound for each of the following steps, one at a time, until he had mastered that step (Wolf, Riseley, and Mess, 1964):

(1) Picking up the glasses
(2) Holding the glasses
(3) Carrying the glasses
(4) Bringing the frames close to his face
(5) Wearing the frames in the proper manner
(6) Eventually: wearing the glasses up to 12 hours with further training, which included weaning from the clicking sound

- For teaching complex tasks, (e.g., learning to drive a car with a standard transmission).

21. *Stress-inoculation training (directive)* (Meichenbaum, 1977). Stress-inoculation involves three separate phases: (a) exploring and discussing the orgin of emotions and how the person responds to the stress (e.g., how the client gets mad, types of problems and when they occur); (b) exploring different ways to cope (e.g., self-talk, deep breathing) and rehearsing these new coping skills; (c) creating an experiment to test the new coping skills under "controlled" stressful conditions. Here is an example of how to use stress inocculation with anger (adapted from Meichenbaum, 1977):

    - Education phase: The client is given the opportunity to explore the anger and find its origin. Three micro steps are: (a) find the cognition associated with anger; (b) explore the somatic-affect, how the body makes changes during the anger (e.g., tension, physiology); (c) determine the behavioral consequences—withdrawal or escalation—and, if escalation, does expression remain verbal or become physical.

    - Rehearsal phase: Clients are introduced to a variety of coping techniques that involve client actions (time out) and/or cognitive actions (self-instruction dialogue). Once clients have the information and the skills, they practice.

    - Application Training: Under a controlled setting, the client is exposed gradually to the stress so they can test out new skills under the stress that causes them the difficulty. The clients are asked to imagine anger-causing situations and to address situations with new skills.

22. *Systemic desensitization (directive)* (Wolpe, 1990). This is an imagery technique for gradually overcoming fearful responses to anxiety-producing events; it is considered "a subprocedure of reciprocal inhibition" (Schwitzgebel and Kolb, 1974) (See counterconditioning.) In this procedure, the client is first assisted to relax physically by any appropriate method (guided imagery, music, hypnosis). Then the client is asked to imagine a low-anxiety item from a prepared list, maintaining the focus while remaining calm, until no more anxiety is felt. Then the counselor guides the client to imagine a more stressful scene, repeating the procedure step by step until the worst item on the list is responded to with calm rather than fear. Example:

- A very shy teenage boy who would like to date, with the counselor's help, makes a list of items to be used for imagery, beginning with the least anxiety-provoking and moving to the worst:

  (1) Talking about schoolwork with a girl he is *not* interested in.

  (2) Talking about schoolwork with a girl he *is* interested in.

  (3) Walking from school with a group of three, including a girl.

  (4) Standing in line at a donut shop with a group of three, including a girl.

  (5) Having a conversation at the table with the same group of three.

  (6) Talking about schoolwork with a girl while alone at the donut shop.

  (7) Telling the girl one thing he appreciates about her.

  (8) Asking to meet the girl at the donut shop tomorrow.

  (9) Asking the girl for a date over the telephone.

  (10) In person, asking the girl for a date.

Summary

Behavioral approaches to counseling vary in their basic philosophy. On one end of the continuum is B. F. Skinner's external conditioning, while on the other is Julian Rotter's internal locus of control. The techniques and methods reviewed in this chapter emphasize one or the other pole to varying degrees.

Many educational institutions, remedial workshops, day care centers and other community facilities utilize behavioral methods regularly. Because of the popular applications of behavior therapy, I believe it is worthwhile for human services counselors to be familiar with the techniques and have them in their toolbox. The information from this chapter is to be used with other theories, so that you to have a wide selection of techniques for your individual counseling orientation.

References

Bandura, A. (1969). *Principles of behavior modification.* New York: Holt, Rinehart, and Winston.

Bandura, A. (1977). *Social learning theory.* Englewood Cliffs, NJ: Prentice-Hall.

Corey, G. (1995). *Theory and practice of group counseling* (4th ed.). Pacific Grove, CA: Brooks/Cole.

Corey, G. (1996). *Theory and practice of counseling and psychotherapy* (5th ed.). Pacific Grove, CA: Brooks/Cole.

Egan, G. (1990). *The skilled helper: A systematic approach to effective helping.* Pacific Grove, CA: Brooks/Cole.

Goldfried, M. R. (1971). Systematic desensitization as training in self-control. *Journal of Consulting and Clinical Psychology 37:* 228–34.

Goldfried, M. R., Davison, G. C. (1976). *Clinical behavior therapy.* New York: Holt, Rinehart, and Winston.

Howatt, W. A. (1995) *Counselling for paraprofessionals: Formulating your eclectic approach.* Nova Scotia: Nova Scotia Community College Press.

Hunt, M. (1993). *The story of psychology.* New York: Doubleday.

Lazarus, A. A. (1971). *Behavior therapy and beyond.* New York: McGraw-Hill.

Lazarus, A. A. (1989). *The practice of multimodel-therapy.* Baltimore: Johns Hopkins University Press.

Meichenbaum, D. (1977). *Cognitive behavior modification: An integrative approach.* New York: Plenum Press.

Mischel, W. (1976). *An introduction to personality.* New York: Holt, Rinehart, and Winston.

Mowrer, O. H. (1947). On the dual nature of learning: A reinterpretation of "conditioning" and "problem solving." *Harvard Educational Review 17:*102–148.

Murdock, M. (1987). *Spinning inward: Using guided imagery with children for learning creativity and relaxation.* Boston: Shambhala.

Pardeck, J. T. (1994, Summer). Using literature to help adolescents cope with problems. *Adolescence 29:*114. San Diego: Libra.

Rachlin, H. (1970). *Introduction to modern behaviorism.* San Francisco: W. H. Freeman.

Rathus, S. A. (1990). *Psychology* (4th ed.). Orlando: Holt, Rinehart, and Winston.

Schwitzgebel, R. K., Kolb, D. A. (1974). *Changing human behavior: Principles of planned intervention.* New York: McGraw-Hill.

Skinner, B. F. (1971). *Beyond freedom and dignity.* New York: Knopf.

Todd, J., Bohart, A. C. (1994). *Foundations of clinical and counseling psychology.* New York: HarperCollins.

Watson, D. L., Tharp, R. G. (1993). *Self-directed behavior: Self-modification for personal adjustment* (6th ed.). Pacific Grove, CA: Brooks/Cole.

Wilson, G. T. (1989). Behavior therapy. In R. J. Corsini, D. Wedding (eds.), *Current psychotherapies* (4th ed.). Itasca, IL: F. E. Peacock.

Wolf, M., Riseley, T., Mess, H. (1964). Application of operant conditioning procedures and the behavior problems of an autistic child. *Behavior Research and Therapy 31:*305–312.

Wolpe, J. (1990). *The practice of behavior therapy* (4th ed.). Elmsford, NY: Pergamon Press.

Zimbardo, P. G. (1985). *Psychology and life* (12th ed.). Glenview, IL: Scott, Foresman.

Chapter 19

Family Systems Therapy: Virginia Satir and Others

We can make the family a real place for developing real people.

Satir, 1988

HISTORY

Out of the ashes of World War II, families began to rebuild themselves. This sudden reunion of family units created numerous issues that drew attention to the uniqueness of family problems, and family systems therapy was born. In the 1950s, this fundamental therapy began to receive some recognition and then gain popularity in the counseling world. Today, representatives from every area of the behavioral sciences and professional disciplines are involved with, and support, a better comprehension of all facets of family functioning (Goldenberg and Goldenberg, 1996).

Because there are a number of distinct counseling models associated with family systems, this chapter is organized somewhat differently from those that preceded it. After a general introduction, the main ideas and techniques will be presented separately for the following models: human validation therapy, structural family therapy, feminist family systems therapy, solution-oriented (or solution-focused brief) therapy, and the Bowen approach to therapy.

For more information about family systems therapy and its various models, contact the American Association for Marriage and Family Therapy, 1100 17th Street NW, 10th Floor, Washington DC 20336-4601.

INTRODUCTION

Family systems therapy asserts that in order to understand an individual the person must be seen in relation to family interactions. Many times the client's presenting problems (biological, psychological, or social) are symptoms of a family structure that is not functioning at its full potential. Corey (1996) goes on to explain that a client's behavior may:

- Serve as a function or purpose in the family

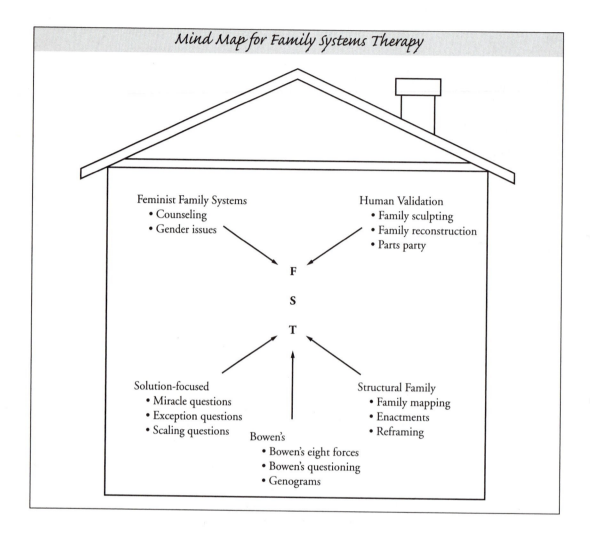

Mind Map for Family Systems Therapy

Feminist Family Systems
- Counseling
- Gender issues

Human Validation
- Family sculpting
- Family reconstruction
- Parts party

F

S

T

Solution-focused
- Miracle questions
- Exception questions
- Scaling questions

Bowen's
- Bowen's eight forces
- Bowen's questioning
- Genograms

Structural Family
- Family mapping
- Enactments
- Reframing

- Be a function of the family's inability to operate productively, especially during developmental transitions

- Be a symptom of dysfunctional patterns handed down across generations

Therefore, from a family systems paradigm, it is not possible to evaluate one person's concerns without an assessment of relationships to family and/or other systems (e.g., work or community) (Goldenberg and Goldenberg, 1996).

There are several differences between systemic and individual approaches to counseling. From the systemic perspective, symptoms of the entire family may be manifested by one family member. Symptoms are seen as indicators of dysfunction within the family, and only dysfunctional patterns are thought to be passed down through several generations. Thus, an individual's level of functioning may be a manifestation of the way in which the family is functioning. Goldenberg and Gold-

Figure 19-1.
Two-
generation
genogram.
(*Note:* With
your clients
I recommend
a three-
generation
genogram.)

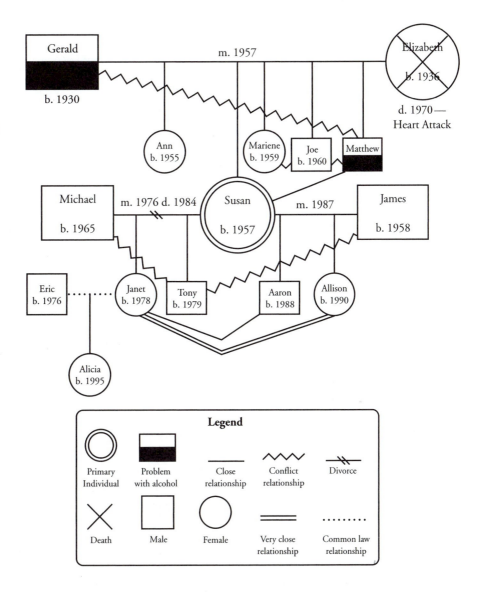

enberg (1996) assert it is possible for a person to be suffering from a symptom or disorder that exists independent of the family structure; it may exist in the community,
or in another system in the individual's life. However, a symptom always affects
other members of the family.

The systemic therapist may use a *genogram* (Figure 19–1) to explore the physical makeup of the family structure (e.g., relationship conflicts). All family members
create their own genogram to reflect the way they see the family system as functioning. From these genograms, the family is able to better understand how each member views the family as a whole. Family interaction is explored in relation to the in-

dividual's symptoms. The systemic therapist explores generational rules, plus cultural and gender focuses, in the family as well as the community or other larger systems in which the person is operating. Satir (1988) proposes that the process of change occurs from within the person as well as systemically.

On the other hand, individual therapists tend to focus on individual assessment and diagnostic tools such as the *Diagnostic and Statistical Manual of Mental Disorders (DSM-IV)*. The plan of action for therapy tends to focus almost exclusively on the client. This orientation explores the causes and purposes, as well as the cognitive, social, emotional, and behavioral processes, involved in the dysfunction of a client. The therapist's focus must be oriented toward the individual's experiences and perspectives, with the goal of helping the person learn to cope with perceived concerns more effectively.

HUMAN VALIDATION THERAPY
Main Ideas

Virginia Satir (1988) explains she developed the model for the human validation process with the intention of emphasizing communication, emotional experiencing, spontaneity, creativity, self-disclosure, and risk taking. This process focused on building self-esteem, negotiating family roles and rules, and developing congruent family communication patterns. Satir identifies three goals of family therapy:

1. All family members are to be honest.
2. All decisions are to be made through negotiation.
3. Each family member is allowed to be unique and differences are to be positively acknowledged so that each person can grow.

In functional families, each member is allowed to live a separate life as well as a shared life within the family. Human validation therapy promotes families working together, so that when they come into conflict it can be looked upon as an opportunity for growth rather than a roadblock. Satir (1988) explains that a dysfunctional family may be identified by closed communication, poor self-esteem of individual family members, and strict rules that set the family up for failure.

Satir (1988) asserts that children are born into families with pre-existing rules. The most important rules, according to Satir, are communication rules involving "who says what to whom under what conditions." Some of these rules can appear impossible to live by, so some individuals in the family may feel they have no choice but to fight them. Human validation principles (Satir, 1988) make no attempt to dismiss the family rules that are thought to cause conflict. Consider the case of a teen who stays out late on a school night. The parents and teen are in conflict, because the teen believes the parents are too strict. Using human validation, the issue of power is not the focus; rather, it is the rationale of coming home early on a school night in order to assure the student's immediate success in school and later career potential.

According to Satir, there are four defensive stances from which people choose when faced with a stressful situation:

- *Placating:* these family members sacrifice themselves to please others
- *Blaming:* these family members sacrifice others to maintain their view of themselves
- Being *super-reasonable:* these family members strive for complete control over themselves, others, and their own environment
- Being *irrelevant:* these family members create a pattern of distractions with the mistaken belief that hurt, pain, or stress will diminish

Satir teaches that when we function exclusively from one of the preceding defensive patterns we are not being healthy and are disrupting the family system. A healthy person would not sacrifice self to a single coping mechanism when dealing with stress. Instead, the healthy individual would transform it into a challenge that is met in a useful way. The words of healthy people match their inner experience, and they are able to make direct and clear statements; they are, as Satir would say, *congruent.*

According to human validation theory, family interactions are influenced by the difficult roles that family members play. Satir (1988) confirms that the roles parents play in the lives of their children are especially important, because children rely on their parents to help them to survive.

According to Satir, when a child is born, adopted, or otherwise introduced into the parents' relationship, the resulting triadic relationship can be dysfunctional for everyone if there are rigid, pre-existing rules to follow and roles to adhere to. However, there is also the possibility of parents forming a nurturing triad with each child, if their rules and family roles are flexible.

There are many techniques supporting human validation theory. Regardless of the counselor's orientation and skill level, an important part is personal congruence of the counselor. Therapists are best conceived as facilitators in charge of the therapeutic process; they do not have the task of making change happen, or curing individuals. Corey (1996) continues that the role of the counselor is to:

- Create a setting in which people can risk looking clearly and objectively at themselves and their actions
- Assist family members in building self-esteem, which helps clients identify their assets
- Take the family history and note past achievements
- Decrease threats by setting boundaries and reducing the need for defenses
- Show that pain and the forbidden are acceptable areas to explore
- Use certain techniques to help restore the client's feelings of accountability
- Help family members see how past models influence their current expectations and behavior and help bring about change in these expectations
- Delineate roles and functions
- Complete gaps in communication and interpret messages
- Point out significant discrepancies in communication
- Identify nonverbal communication

Techniques

All of the following techniques are useful and powerful; however, we strongly suggest the human services counselor be well trained and supervised in them before trying to utilize them.

1. *Family sculpting.* The counselor positions each family member in relation to the whole, often using nonverbal communication stances to demonstrate how family members interact and communicate with each other (Sherman and Fredman, 1986). Example:

   • Counselor positions the father in a raised-arm stance, because he is always lecturing his children. The children are positioned kneeling with their heads bowed. It allows the father to experience the full impact of his lecturing to his children, and to see a picture of how the children perceive themselves during his lecture. This technique relies on the premise "A picture is worth a thousand words."

2. *Family reconstruction* (Satir and Baldwin, 1983). Takes clients through different stages of their lives with three goals in mind: (1) to enable family members to identify the roots of old learning, (2) to help them formulate a more realistic picture of their parents, and (3) to assist them in discovering their own unique personality. Example:

   • A client is asked to develop a chronological history of himself and his family starting as far back as he can remember. The counselor then takes these three points as paradigms to assist the client in reconstructing a more helpful interpretation of the parts, so that the client can feel better about the present. Look again at the genogram presented earlier (Figure 19–1). It is recommended that the client go back at least three generations in order to see his parents as people and to get to know some of his family history.

3. *Parts parties.* Each person is seen as a system of parts, both positive and negative. People often distort, deny, or disown parts that are less useful in adolescent and adult life but that served the younger child's need for survival. Example:

   • A simple parts party with couples might invite each partner to list six of each other's characteristics as represented in the form of well-known public figures. This is done to help partners realize that some of their characteristics may be incompatible while others will live harmoniously with one another. For instance, one partner may choose Abraham Lincoln for integrity, Robert Redford for sexiness, and Meryl Streep for sophistication. The other partner picks Archie Bunker for grouchiness, Jane Alexander for intelligence, and Martin Luther King, Jr., for strength. The couple then begins to play the characters. If Meryl Streep and Archie Bunker meet, it is unlikely they will get along; however, if Jane Alexander and Abraham Lincoln meet, they will most likely get along. The purpose is to allow the participants to see and experience how some aspects of their personalities clash, and how some conflict is inevitable. To get out of conflict, the individuals must learn to rely on a different part of their personality.

Figure 19–2.
Family supra-
system with
subsystems.

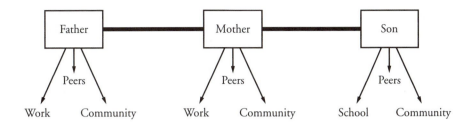

STRUCTURAL FAMILY THERAPY
Main Ideas

Structural family therapy was, in part, created in the 1960s by Salvador Minuchin. This orientation is based on the idea that symptoms of dysfunction are produced from structural failings within the family organization (Sherman and Fredman, 1986). There are invisible sets of rules or functional demands organizing the way members of a family relate to one another. These rules comprise the family structure; for example, the father states a rule that the son is to do as father says. The aim of structural techniques is to "bring about change [and]. . . . reorganize the system by getting members to move from one place to another, from one role to another" (Sherman and Fredman, 1986). Therefore, the goals of structural family therapy are to reduce symptoms of dysfunction and to bring about structural change within the system by modifying the family's transactional rules, and developing more appropriate boundaries.

Goldenberg and Goldenberg (1996) confirm that families are basic human systems made up of subsystems, with each part existing at all times (Figure 19–2). Each family member is part of the family suprasystem, while having his or her own subsystems. When family members get involved without permission in the other member's subsystems, there is usually conflict.

When members of a family subsystem intrude on, or try to take over, a subsystem in which they do not belong, there is usually some type of structural difficulty (e.g., when parents become involved in their children's relationships outside the family unit). *Boundaries* determine the amount of contact one individual has with another; they are emotional barriers that protect and enhance the integrity of individuals, subsystems, and families. These boundaries are best understood if seen as part of a continuum ranging from rigid to diffuse. *Rigid boundaries* lead to impermeable barriers between the subsystems and among subsystems outside the family. In the middle of the continuum are healthy boundaries that effectively blend the rigid and diffuse characteristics. *Diffuse boundaries* allow members of a family to maintain their own identities without threatening the existence of the entire family system.

The role of the counselor in structural family therapy involves three interactive functions: (1) joining the family in a position of leadership, (2) mapping its underlying structure, and (3) intervening in ways designed to transform an ineffective structure. The following are techniques to be utilized in the support of this role.

Techniques

1. *Family mapping.* To implement this technique, a variety of maps are drawn identifying boundaries within a family; boundaries are either rigid, diffuse, or healthy. These boundaries are examined and successfully resolved so that the family may move forward to the next stage of development. The maps depict a family's life cycle, including interpersonal relationships within the family, and can be very effective in therapy sessions. Example:

   - As in Figure 19–2, each family member maps out his or her subsystems and explores the communication barriers between individuals, as well as areas of concern relating to perceived violations of subsystems.

2. *Enactments.* This technique is most effective if transactions between family members are kept in the present moment. Example:

   - The counselor asks the client to act out a situation that is causing conflict at home. For instance, a teenager may be questioning an eleven o'clock curfew and thus angering his father, who feels the curfew is perfectly legitimate. This allows the counselor to experience the breakdown in communication and the resultant subsystem barriers that have been damaged or intruded upon.

3. *Reframing.* The counselor reframes in order to shed new light on an old problem, thus providing a different interpretation of a particular situation. This allows the presenting problem to be explored in ways that enable the family to see the complaint from many different angles. Through reframing, it becomes possible to grasp the underlying family structure that is contributing to an individual's problem. Example:

   - A child acts out in school. His parents are in the midst of a divorce. Every time the child acts out, the parents come together to solve the problem. The counselor should then point out that the parents are not fighting when they are brought together for a common goal regarding the child. Because the child recognizes this, she will continue to act out until the presenting problem is addressed.

FEMINIST FAMILY SYSTEMS THERAPY
Main Ideas

Corey (1996) reports that there is no singular, unified feminist family systems therapy; however, feminists do share some common ideologies:

1. A belief that patriarchy is alive and sick in sociopolitical life and in the life of the family

2. A realization that the normal family has not been so normal or wonderful for mothers and clearly reflects the discrimination against women evident in world systems beyond the family

3. A commitment to reforming family and society in ways that fully empower and enfranchise women economically, socially and politically

4. Therapeutic processes that include a positive attitude toward women, social analysis, explicit consideration of gender issues, and treating the personal as political

Gilligan (1982) asserts that, with feminist theory, the primary objective for psychotherapy in the family unit is to challenge the pathological ideals of the subservience of women within families and society. Gilligan notes that developmental psychologists ranging from Freud and Erikson to Piaget and Kohlberg had all based their theories on male populations, assuming either that women's growth was similar to that of men, or that women were insignificant exceptions to the male norm. Feminist theories allow families the opportunity to address issues of power within the family and in its surrounding systems. Gender-based roles and rules are challenged while systemic problems—family violence, cultural discrimination, ageism, poverty, race, and class, as well as discrimination against gay men and lesbians—are targeted for change.

Techniques

There is no specific set of feminist techniques, because of the diversity of feminist theory. However, Corey (1996) points out some commonalities shared by feminist theorists.

Counseling is to be conducted with a conscious purpose and:
- a positive attitude toward women
- an orientation that values that which is considered feminine or nurturing in society and social interactions
- a willingness to confront patriarchal processes and reinvolve fathers in family life
- empowering women while supporting egalitarian families

Six ways in which gender issues might be introduced into family therapy:
1. Define the problem in such a way as to include the dimensions of power and gender.
2. Introduce and discuss in therapy gender issues such as money, power, equity, flexibility, options, housekeeping, and childcare.
3. Make connections for the family between gender issues in the family and those in the wider social system.
4. Challenge stereotypic behaviors, attitudes, and expectations.
5. Discuss the differing impact of divorce on women and men.
6. Raise gender issues in relation to the family of origin.

Example:
- Counseling is conducted with a conscious purpose and a focus on gender issues. A problem is stated using behavioral language. Each family member is given the opportunity to speak, and listen, without getting defensive. Cormier and Hackney (1987) believe every angle of the problem should be examined. From a gender and power perspective, for instance, a woman

might have very low self-esteem and not understand why until she examines her relationship with her husband. He stereotypically insists on running the household and driving her everywhere she needs to go. He feels this is expected of him because his father behaved in the same way with his mother. He does not understand that it is his domineering behavior that is leading his wife to file for divorce.

SOLUTION-ORIENTED THERAPY
Main Ideas

William O'Hanlon, Michelle Weiner-Davis, and Steve deShazer began developing solution-oriented therapy in the late 1970s. Goldenberg and Goldenberg (1996) state that the premise of solution-oriented therapy is that clients, with the aid of the counselor, try to discover solutions that will lead them to reach goals they have set for themselves.

Goldenberg and Goldenberg (1996) find many people bring their own interpretations of events into therapy. In a problem-oriented state, some individuals use these narratives to justify lack of change in their lives, or as an excuse by blaming life for moving them further and further from their goals. It is the job of the therapist to find the good or the positive in these stories; for "it is in these stories of life worth living that the power of problems is deconstructed and new solutions are manifested and made possible."

Solution-focused therapy grew out of the need to find solutions as quickly as possible so the family can face forward as a happy and loving unit. Solution-focused therapy does not look at the source of problems within families; rather, the therapist aids the family in describing the presenting situation. Having identified the family's goals, the counselor helps the family to find solutions that are congruent with the goals. Clients are encouraged to talk about these solutions rather than listing facts about their troubled lives.

The underlying assumption is that clients have within themselves the capacity to solve their own problems and the therapist's role is merely to bring this knowledge to the forefront and guide the solution process. Metaphorically, clients' complaints are like locks on doors that could open to a more satisfactory life, if only the key could be found. "Often, time is wasted and frustration heightened. . . trying to discover why the lock is in the way or why the door won't open, when the family should be looking for the key" (Goldenberg and Goldenberg, 1996). The solution-focused family therapist need only aid the family in finding a skeleton key that will fit a variety of locks. Then the family has the tools to solve several problems rather than just one. The goal is that the family will experience that problems can disappear quickly and there is often no need to dwell on the past.

Techniques

Solution-focused therapists counter the problem-oriented state of the client by initiating optimistic *solution conversations* that are built around the clients goals and beliefs and work toward goals that are attainable. Some methods used to facilitate these conversations include:

1. *Miracle question.* If a miracle happened overnight and your problem was solved, how would you know it was solved? What would be different?

2. *Exception-finding question.* Direct clients to a time in their lives when they did not suffer from their problem. These questions focus on a time when there were exceptions to the rules.

This exploration reminds the client that the problem has not existed forever, provides the opportunity for the exploration of the client's supports and resources, and may lead to a solution. Using these questions, solution-focused therapists build on occasions when the client could control the problem (Goldenberg and Goldenberg, 1996). Example:

- Last Saturday Jack was home for his twelve o'clock curfew. What would have to happen to get him home on time this Saturday?

3. *Scaling questions.* These are best utilized when dealing with affect, communication, or moods. Example:

- On a scale of zero to ten, how do you feel about your brother's behavior now?

BOWEN APPROACH TO THERAPY
Main Ideas

One of Bowen's main tenets was the importance of theory as a blueprint for therapists. In his search for disciplined theoretical approaches, Murray Bowen became interested in the relationships between mothers and their schizophrenic children. It was during these studies that he found relationships between mothers and children to be more emotionally intense than he had hypothesized. Bowen also discovered that emotional relationships throughout the family, including those with father and siblings, played key roles in perpetuating family problems.

Thus, Bowen established the entire family as an emotional unit made up of members who are connected and influenced by each other. The underlying premise of Bowen's family systems theory is that family members are always in a struggle between emotional togetherness and emotional distance from each other.

Bowen's (1978) concept of chronic anxiety is always present in everyday life. This anxiety is passed down from generation to generation and is experienced when an individual tries to separate the self from the rest of the family. According to Bowen, this chronic anxiety is the foundation of all symptomology; its only antidote is resolution through *differentiation,* the process by which individuals learn to chart their own direction in life rather than perpetually following the guidelines of others.

Bowen's theory of family systems therapy rests on the idea that there are forces within the family that allow for togetherness and there are forces in opposition that lead to individuation. They note the following eight forces influencing family functioning.

1. *Differentiation of self.* This occurs when individuals are able to avoid having their behavior automatically driven by their emotions. An individual functioning at an ideal level would not be emotionally detached—or completely objective. Such individuals would be able to tell the difference between their cognitive

processes and their affective processes without losing their capacity for spontaneous emotional expression. Some people cannot tell the difference between the
cognitive and the affective. Their behaviors are automatically driven by their
emotions. This is known as a *fusion* of thought and action (without any differentiation of self).

2. *Triangles.* The triangle is a basic building block in a family's emotional system.
   When anxiety is low, two people may have an easygoing back and forth exchange of feelings. But, if one or both individuals become anxious, they may
   reach out to involve a third person. Their anxiety is then diffused within the triangle, which relieves symptoms but masks the originating problem. It is possible
   for this triangulation to move beyond the immediate family to involve other social systems.

3. *Nuclear family emotional system.* If two undifferentiated people marry, it is possible that, as a couple, they will become as fused to each other as they once were
   to their families of origin. Their resulting nuclear family emotional system will
   be unstable. The more fused the family is, the higher their anxiety levels will be
   —and the greater the likelihood is that they will eventually try to distance themselves from each other in a quest for differentiation of self. This can lead to marital difficulties.

4. *Family projection process.* Establishes that parents focus on the most immature of
   their children to project their own low levels of differentiation on. These children tend to develop a stronger fusion to the family than their siblings; therefore, they are more vulnerable to stress in the family.

5. *Emotional cutoff.* Children who are enmeshed in the projection process will try
   to protect themselves from complete fusion with their families. They will do this
   by running away, ceasing to speak with their parents, or by actually breaking
   contact completely. Emotional cutoffs occur more often within families who
   have high levels of anxiety and emotional dependance.

6. *Multi-generational transmission process.* For the multi-generational transmission
   process to exist, two previously discussed elements are crucial: (1) the selection
   of a spouse with a similar differentiation level, and (2) the presence of the family projection process that instills low levels of self differentiation generation after generation. As each generation produces individuals with progressively
   poorer differentiation, family members become increasingly vulnerable to anxiety and fusion.

7. *Sibling position.* Sibling position is important to the nuclear family emotional
   process, because children develop specific personality characteristics according
   to birth order. When they marry, they maintain these characteristics. Therefore,
   the more a marriage duplicates an individual's sibling place in childhood, the
   more likely it is to succeed. Thus, the youngest should marry an older child, and
   the firstborn should marry a secondborn.

8. *Societal regression.* In this theory, Bowen suggests that society, like a family, has
   within it forces acting toward differentiation as well as those opposed to it. For

example, under conditions of social unrest and high anxiety, such as population growth, there can be a surge of togetherness that erodes forces driving toward differentiation of intellect and emotion. The result, according to Bowen, will be increased anxiety levels and a perpetuation of the problem.

Techniques

Bowen's theories are always governed by two basic goals: (1) the reduction of anxiety and relief from symptoms, and (2) an increase in the level of self-differentiation in the client. Bowen, when working with parents and their symptomatic child, would try to show the parents that the problem was not with the child, but between *them* as the family's main emotional system.

1. *Questioning.* Act as a coach, using low-key direct questioning to help clarify emotional responses between two people. The goal of questioning is to help clients differentiate their thoughts from their emotions by using the eight forces that influence family functioning. Example:

   - What are you thinking about when you become so emotional and behave that way? How do you feel about yourself when you behave that way?

2. *Genograms.* Bowen used genograms to gather a detailed history of the client and to help bring into awareness ways in which the history may bear on present day concerns. (*Note:* For a large visual illustration, look again at the two-generation genogram in Figure 19–1.) When you are doing a genogram with the client, do three generations so you can obtain plentiful family history for establishing any chronic family behaviors (e.g., problems with alcohol).

Summary

In today's world we have come to realize that family breakups because of divorce are commonplace. Counselors need to become aware of how important it is to be able to deal with the family as well as with individuals. This chapter introduced the human services counselor to five models of family therapy, providing some insight into the rationale and techniques of the various models. I strongly recommend that, without both intensive training and experience in family therapies, human services counselors be conscious of their limitations and ready to make appropriate referrals.

References

Bowen, M. (1978). *Family therapy in clinical practice.* New York: Aronson.

Corey, G. (1996). *Theory and practice of counseling and psychotherapy* (5th ed.). Pacific Grove, CA: Brooks/Cole.

Cormier, L. S., Hackey, H. (1987). *The professional counselor: A process guide to helping.* Englewood Cliffs, NJ: Prentice-Hall.

Davis, K. (1996). *Families: A handbook of concepts and techniques for the helping professional.* Pacific Grove, CA: Brooks/Cole.

Gilligan, C. (1982). *In a different voice.* Cambridge MA: Harvard University Press.

Goldenberg, I., Goldenberg, H. (1996). *Family therapy: An overview.* Pacific Grove, CA: Brooks/Cole.

Haley, J. (1973). *Uncommon therapy: The psychiatric techniques of Milton H. Erikson, M.D.* New York: Norton.

Hanna, S., Brown, J. (1995). *The practice of family therapy: Key elements across models.* Pacific Grove, CA: Brooks/Cole.

Minuchin, S., Fishman, H. C. (1981). *Family therapy techniques.* Cambridge, MA: Harvard University Press.

Satir, V. (1988). *The new peoplemaking.* Mountain View, CA: Science and Behavior.

Satir, V., Baldwin, M. (1983). *Satir step by step: A guide to creating change in families.* Mountain View, CA: Science and Behavior.

Sherman, R., Fredman, N. (1986). *Handbook of structured techniques in marriage and family therapy.* New York: Brunner/Mazel.

Part III

First Aid Counseling

Chapter 20

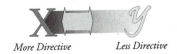

More Directive   Less Directive

Introduction to Crisis Management

Unless the person obtains relief, the crisis has the potential to cause severe affective, cognitive, and behavioral malfunctioning.

Gilliland and James, 1996

MAIN IDEAS

This chapter is designed to provide an introductory level of knowledge and skills for handling potential crises that may arise. It includes interventions for suicide, abuse, addictions, HIV/AIDS, grief, eating disorders, and stress. The purpose of this chapter is to provide you with a framework from which to operate when dealing with crisis. Beginning with a crisis model is helpful when you are first learning to do interventions. From this perspective, you can help the client build a skill set that will lead to healthier choices. It also provides a framework around which you can establish your own counseling orientation.

It is important to realize that a crisis can also be viewed as an opportunity because it is usually the point at which most people seek help (Gilliland and James, 1996). The following three definitions illustrate various aspects of a crisis. Slaby (1985) states crises are *normal*, in the sense that people who experience a crisis feel overwhelmed and that no one can predict whether an event will trigger a crisis. For example, a firefighter of fifteen years rescues a small child from a burning building. Does the firefighter experience this event as a crisis?

France (1990) describes crisis as *personal*. What affects one person may not affect the next person. The perception of the individual involved in the incident will determine the impact of the event on that person. Suppose the firefighter noticed that the small child resembled his daughter. Might this affect whether he experiences a crisis?

Aguilera (1994) defines crisis as occurring "when a person faces *obstacles to important life goals* [italics added] that are, for a time, insurmountable through the utilization of customary methods of problem solving." The firefighter disobeyed a direct order in returning to the burning building and is suspended as a result. Is this experienced as a crisis?

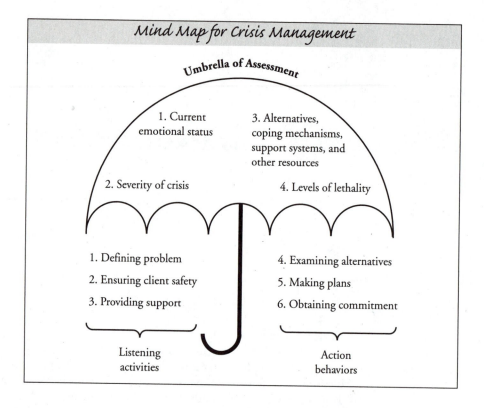

Crisis intervention can be defined as "any beneficial behaviors used to assist a person who perceives his/her situation to be catastrophic to function more effectively and return to a state of homeostasis" (Davis and Sandoval, 1991).

Gilliland and James (1996) list characteristics a human services counselor will need to be an effective crisis worker:

- Knowledgeable
- Adaptable / flexible
- Resourceful
- Attentive / perceptive
- Therapeutically congruent
- Calm, poised / prepared [to work effectively] with diverse and complex issues as they arise
- Creative
- Physically, emotionally, and mentally present for the client
- Energetic
- Optimistic

- Able to . . . assess and synthesize information [quickly and accurately]
- Multiculturally sensitive

MULTICULTURAL CONSIDERATIONS

Ignoring the many multicultural perspectives that pervade our society can result in a failure to understand the world as the client sees it. It can influence the counselor to make interpretations, judgments, and conclusions that may be based on irrational assumptions or biases. Being multiculturally effective requires that counselors know themselves and be aware of the concepts about which they may have preconceived notions. These may include: definitions of "normal" behavior; the importance of independence; and forms of prejudice such as ageism, racism, religionism, sexism, ethnocentrism (Gilliland and James, 1996).

An effective professional will seek knowledge to dispel biases. This knowledge may come from researched information or from the client (Martin and Moore, 1995). Using this knowledge, the counselor can assess the client's problem more accurately and collaborate to formulate a successful intervention and treatment plan. The counselor therefore does not try to impose values on the client. Sue (1992) states that multiculturally sensitive counselors "use methods and strategies and define goals consistent with the life experiences and cultural values of the client" (Gilliland and James, 1996).

ETHICAL CONSIDERATIONS

During a crisis emotions often run high, and through it all the counselor must be aware of ethical considerations. It may be necessary to review with the client the basic rules of confidentiality and its limitations. Because each different geographic region has its own laws, statutes, and regulations regarding confidentiality, counselors *should not assume* they know these laws; they should become knowledgeable about the ethical standards and regulations for their specific location (France, 1990). The client must be notified of all obligations the counselor must abide by as well as which disclosures and behaviors must be reported to authorities. It is important to remember that ethical codes are as much for the protection of the therapist as they are for the client (Martin and Moore, 1995).

TECHNIQUES

In their book *Crisis Intervention Strategies,* Gilliland and James (1996) introduced the six-step model of crisis intervention that follows shortly. No matter what aspect of "first aid" you are addressing (e.g., addictions, suicide), I suggest you use the six-step model in combination with the skills you obtain from the appropriate chapter of this book. The six-step model will provide you with a frame of reference to help ensure the client remains safe.

In the same work, Gilliland and James also provide a quick assessment model called the *triage assessment system* (1993). This allows a quick evaluation of the client's current functioning in three domains—affective, cognitive, and behavioral—so that you can decide how directive you need to be to help move the client out of crisis. The triage assessment system would be used in step 1 (defining the problem) of the six-step model of crisis intervention.

The Triage Assessment System

Affective Domain Scale

The counselor rates the client's degree of expressiveness of three emotional states: anxiety, anger/hostility, and depression or sadness/melancholy. The counselor assigns a rating, on a 10-point scale, with 1 representing stable, appropriate affective functioning, and 10 representing a state of depersonalization (Figure 20–1).

Cognitive Domain Scale

During a crisis, a client will often fall prey to various maladaptive thoughts and beliefs. Thus, three types of cognitions—transgression, threat, and loss—are assessed in four areas of the client's life: (1) physical needs, (2) psychological needs (identity, emotional health), (3) social relationships, and (4) moral/spiritual needs. On each scale, the counselor again assigns a rating between 1 and 10. One represents normal problem-solving and decision-making abilities, while 10 represents an inability to concentrate on anything but the crisis. The client's perception of the crisis may differ from reality (Figure 20–2).

Behavioral Domain Scale

Clients' behavior can also be described along three dimensions. They may approach, avoid, or become paralyzed and immobile, in response to a crisis. Each of these behaviors is assessed and the counselor assigns a rating between 1 and 10. One repre-

Figure 20–1.
Affective domain (adapted from Gilliland and James, 1993)

Affective Domain
Anxiety; Anger/Hostility/ Depression

1 2 3 4 5 6 7 8 9 10
Stable   Very Unstable

Figure 20–2.
Cognitive domain (adapted from Gilliland and James, 1993)

Cognitive Domain

1 2 3 4 5 6 7 8 9 10
Stable   Very Unstable

	Physical	Psychological	Social	Moral/Spiritual
Transgressions	☐	☐	☐	☐
Threat	☐	☐	☐	☐
Loss	☐	☐	☐	☐

Figure 20–3.

Behavioral domain (adapted from Gilliland and James, 1993)

	Behavioral Domain (check appropriate box)										
	No Impairment			Moderate Impairment				Severe Impairment			
	1	2	3	4	5	6	7	8	9	10	
Approach											
Avoidance											
Immobility											

sents appropriate coping, while 10 represents erratic, unpredictable, often harmful behavior (Figure 20–3).

In scoring, each domain is given a score between 1 and 10, which represents the client's overall functioning in that domain. The ratings of all three scales are then combined to give a final score. A score between 3 and 10 would indicate minimal impairment; between 15 and 20, moderate to high impairment; in the high 20s, severe impairment requiring immediate intervention.

The Six-Step Model of Crisis Intervention

1. *Define the problem.* This requires understanding the problem from the client's perspective. It is of utmost importance because this understanding forms the basis upon which prevention and treatment plans are formulated. It is recommended that the counselor use active listening skills and reflection to achieve this understanding (see Chapter 8). Many times, unless we hear and understand from the client's position, we never really know what the concern is. Examples:

   • Using open-ended questions, the counselor will receive answers that more fully detail and represent the client's concern:

   COUNSELOR: Explain to me the situation as you see it.

   COUNSELOR: Tell me about what is going on in your life.

2. *Ensure client safety.* This involves "minimizing the physical and psychological danger to self and others" (Gilliland and James, 1996). Safety is paramount throughout crisis intervention, and it should be at the forefront of the counselor's mind through every client/counselor encounter. Depending on the client's state of mind, the counselor will need to determine how directive to be. If a client wants to die by taking his own life, the intervention will be very directive and enlist the support of police, family, and others. It is important to understand that, when clients are in crisis, they have trouble with problem solving, and therefore need direction. Examples:

COUNSELOR [directive]: I want you to stay here until I can get you some support.

COUNSELOR [non-directive]: Who can help you with this problem?

In some situations it will be readily apparent the client is at risk to be harmed, by self or another individual. The client's safety can be enhanced by seeing that she is removed from the situation and has someone with her who can assist her or act on her behalf.

3. *Provide support.* This involves communicating to clients that the counselor is present for them and cares about them. The counselor must consistently provide unconditional positive regard for the client as a human being. Example:

   • The counselor's verbal and nonverbal communication is congruent. The counselor actively listens and is attentive.

4. *Examine alternatives.* This involves brainstorming in order to uncover a wide variety of resources and choices available to the client. Example:

   COUNSELOR: Let's go through all of the resources you have available.

   COUNSELOR: Now we know the resources available, exactly what is it you want to happen?

   COUNSELOR: With the resources you have available, what alternatives do you have to obtain the outcome you want?

5. *Make plans.* This will involve collaborating with the client to formulate a realistic, action-oriented strategy that is attainable, given the client's current mental state and capabilities. Example:

   CLIENT: The next time I am feeling frustrated, I will go for a walk. This will get me out doing something, so I don't sit around stewing about Ann. I do not want to think about hurting myself, and if I continue to feel frustrated, I will.

   COUNSELOR: Sounds like a step in the right direction.

6. *Obtain commitment.* This involves having the client commit to definite, positive actions and be responsible for undertaking them. Example:

   • Restate the plan, ensuring that the client takes ownership of the plan. In some instances, the client's verbal agreement is sufficient; however, in situations like suicide intervention a written contract may have greater impact for the client (see Chapter 22).

Acknowledgments

The author is indebted to Burl E. Gilliland and Richard K. James for much of the material included in this chapter. The reader is referred to their *Theories and Strategies in Counseling,* Fourth Edition (Prentice-Hall, 1984), and *Counseling and Psychotherapy,* publishing by Allyn & Bacon in 1998. These works more completely delineate their ideas.

Summary

This chapter provided a definition of crisis, an assessment model for determining how the crisis has affected the client (Triage Assessment), and a crisis intervention model (Six-Step). Using these elements, you can create a frame of reference from which to work when dealing with crises as well as a guide to follow when working with any of the aspects of "first aid."

It is important to note that every person responds differently to a crisis. The way one individual acts and the coping mechanisms he employs is no indication how another person will, or should, react. Every client is a unique individual, and needs to be treated as such.

In closing, no matter what your counseling orientation, I highly recommend you become familiar with Gilliland and James (1996) six-step model. It provides an excellent template to help your client (and you) get through the crisis.

References

Aguilera, D. (1994). *Crisis intervention theory and methodology* (7th ed.). St. Louis: Mosby.

Davis, J., Sandoval, J. (1991). *Suicidal youth: School-based intervention and prevention.* San Francisco: Jossey-Bass.

France, K. (1975). Evaluation of lay volunteer crisis telephone workers. *American Journal of Community Psychology 3*:197–220.

Gilliland B., James, R. (1996). *Crisis intervention strategies* (2nd ed.). Pacific Grove, CA: Brooks/Cole.

Martin, D. G., Moore, A. D. (1995). *First steps in the art of intervention.* Pacific Grove, CA: Brooks/Cole.

McGee, R. F. (1984). Hope: A factor influencing crisis resolution. *Advances in Nursing Science 6* (4):34–44.

Slaby, A. E. (1985). Crisis-oriented therapy. In F .R. Lipton, S. M. Goldfinger (eds.), *Emergency psychiatry at the crossroads: New directions for mental health services.* San Francisco: Jossey-Bass.

Chapter 21

More Directive   Less Directive

Eclectic Verbal Intervention

The best way to defuse a situation is to listen, think, and then ask another question.

Howatt, 1999

MAIN IDEAS

Verbal intervention skills can both prevent the escalation of a client crisis and decrease job stress for the counselor. When faced with a client in crisis, professionals will adapt their own personal orientation to obtain the best possible resolution. When working with an assortment of clients, you may encounter individuals who act out, either verbally or physically. Faced with such a situation, counselors can fall back on guidelines to minimize the situation and guide the client back to stability. Crisis intervention is important in three ways:

1. The individual is more receptive to help during a crisis.

2. The result, with intervention, is a learning experience for the individual.

3. It can save suicidal individuals or those with severe personality disorganization (nervous breakdown). (France, 1990)

It is important to notice early warning signs that indicate a client is escalating toward potential acting-out behavior. The client may become restless, irritable, agitated, nervous, or defensive. This can be an indication that the client is responding to some threat to well-being, real or perceived. Out of fear, the client may create a behavior in an attempt to regain control of the situation (e.g., the client may start verbally to abuse the counselor).

TECHNIQUES

Step 1 *Stay calm.* The professional must maintain composure and stay calm. This can be achieved through self-talk. Statements made to yourself silently can assist you to think clearly and objectively about the situation. The objective of self-talk is to help you defuse the situation by remaining calm and in control. The premise underlying self-talk is that the way you talk to yourself is crucial in determining how you behave, think, and feel. Examples of self-talk (Meichenbaum, 1977):

- I can handle this.
- Just stay calm and take a deep breath.

Step 2   *Be aware of nonverbals.* Adler and Towne (1993) suggest that nonverbal communication is 12.5 times more powerful than words. Professionals should be aware of their own nonverbal behavior, as well as the client's. Egan (1994) explains that clients send clear messages through their nonverbal behaviors (e.g., body posture, voice tone); the professional needs to read these messages without distorting or overinterpreting them.

By recognizing the client's nonverbal behavior, the counselor is better able to interpret . . . what the client is feeling; this helps the professional gain awareness as to whether or not the client's behavior is escalating (Egan, 1994). The following examples of anxiety, anger, and coldness will provide the counselor with some general guidelines for reading a client's nonverbal behavior:

- Anxiety: The client may gesture more, get red in the face, sweat. The anxious individual may also frequently shift in her seat, tap fingers or a foot, or start pacing the floor.
- Anger: The client will frown, tense lips, widen eyes, and thrust the chin and head forward. Anger may escalate into physical acting-out behavior.
- Coldness: The client will be unsmiling, have little or no eye contact, and have a closed posture (arms folded across chest, legs crossed high). If wearing glasses, the client will remove them.

Counselors must also be aware of personal space (proxemics) in a potential crisis situation. Adler and Towne (1993) define *proxemics* as a form of nonverbal communication in which people use space to communicate. The amount of personal space required will vary from client to client. People have different space requirements at different times and in different situations. The distance a person chooses depends upon three things (Adler and Towne, 1993):

- How the individual feels toward the other person at a given time
- The structure of the conversation
- Interpersonal goals

The counselor needs be aware of *distances* in potential crisis situations because it is important to be know how we can validate intimate personal space, which is often the source of the client's discomfort.

- Intimate distance: This begins with skin contact and extends outward to approximately 18 inches. When people allow others to move into their intimate space voluntarily, it is a sign of trust. When someone invades this space without the other person's consent, the person usually feels threatened; thus, the counselor needs to be aware of the client's intimate space. If the counselor "invades the client's space, the client will feel threatened, and may withdraw or act out" (Adler and Towne, 1993).

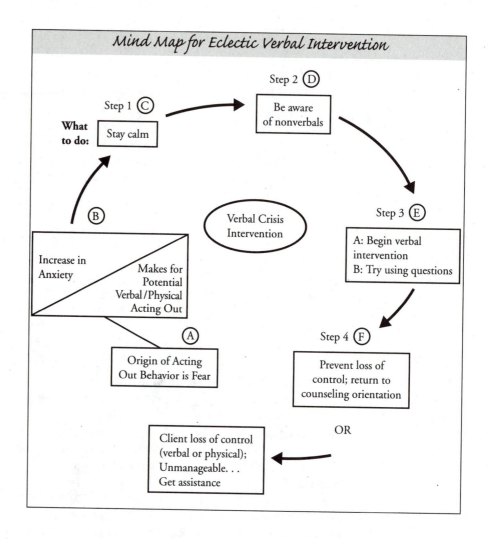

Mind Map for Eclectic Verbal Intervention

Step 1 (C)

What to do: Stay calm

Step 2 (D)
Be aware of nonverbals

(B)
Increase in Anxiety / Makes for Potential Verbal/Physical Acting Out

Verbal Crisis Intervention

Step 3 (E)
A: Begin verbal intervention
B: Try using questions

(A)
Origin of Acting Out Behavior is Fear

Step 4 (F)
Prevent loss of control; return to counseling orientation

OR

Client loss of control (verbal or physical); Unmanageable. . . Get assistance

---

- Personal distance: This ranges from 18 inches to about 4 feet. The far range of personal distance goes from 2.5 to 4 feet. It is appropriate for the professional to be 3 to 4 feet from the client in the counseling session. It is better to err on the side of caution than to invade personal space. If the client begins to display acting-out behavior, it is suggested that the counselor lengthen the space. Counselors must be aware of the culturally accepted personal space zones for the population with whom they are working. In most North American cultures, the accepted space between two people conversing socially is approximately three feet.

Step 3A *Begin verbal intervention.* The human services counselor uses an array of communication skills to minimize the risk of increased acting-out behavior by the client. Counselors will rely on their own communication model, keeping in mind the importance of empathy and active listening skills. Gossen (1993) teaches that, when you are involved in a verbal crisis intervention, the following points should be considered:

- It's not what you say, but how you say it (voice, speed, and volume),
- Avoid rhetorical questions (You know what to do, don't you ?)
- Stick with the facts, avoid personal opinions (don't say, You always act this way!).

When we talk to clients who are upset, we need to be especially mindful that we do not appear to be judging them. I suggest that you avoid using statements in the early stage and that you rely on open-ended questions. The strategy is to buy sufficient time to allow the client to cool down.

Communication skills involved in verbal crisis intervention can be categorized as follows. *Active listening skills* (Egan, 1994) are used to:

- Observe and read the client's nonverbal behavior (e.g., The client, wide-eyed and frowning, says "I'm not angry").
- Hear and understand the client's verbal message. Reflecting the client's feelings or thoughts also demonstrates interest and empathy. The client is encouraged to explain more fully what is happening ("You appear to be upset").
- Develop trust between counselor and client. The client feels heard and understood ("I heard your point—can you review it for me, so I am sure I understand?")

Through the *use of empathy* (Adler and Towne, 1993) the counselor communicates to the client that expressed feelings are understood ("I can see that you are getting frustrated"). For a discussion of the use of *"I" language,* see Chapter 13.

Step 3B *Try using questions.* This section is an adaptation of Gossen's (1993) intervention model called "What are you shooting for?" The counselor asks various questions to minimize the crisis and reduce the chances of it escalating. If the client does not answer the questions, Gossen suggests turing the questions into statements. Example:

- "What do you want?" If there is no response, say "I need you to . . ." (e.g., to calm down).
- "What are you doing?" If there is no response, say "This is what I see [hear] . . ." (e.g., I see you getting upset and I hear you swearing).
- "How is it working out for you? Is it helping you get what you want?" If there is no response, say "It's not working for me. This is what I want from you . . ." (e.g., I want you to calm down or I will have to stop the session).

Again, the use of questions will give the client time to cool down. In my opinion, the question creates a cognitive "bump," meaning it moves the client out of the emotions and into thinking. Whenever possible, we want to help move the client away from the emotions that are fueling the internal conflict (Howatt, 1996).

Step 4 *Prevent loss of control.* When escalation is taking place, the client may not respond to your verbal intervention and you may perceive the situation as deteriorating into a personal attack (the client is verbally abusive or physically threatening). We recommend the human services counselor follow the suggestions set out by Coloroso (1994). You must use your own best judgment if your safety is threatened. Examples:

- Call time out. The counselor should not take abuse from a client ("I see things are getting out of control; lets take a break and then talk").

- Express your own feelings ("I'm feeling a bit overwhelmed. How about taking a break until we both calm down?").

- Stop counseling. If the client continues to threaten a physical attack, get out of the situation immediately and seek help.

Summary

As a human services counselor, you need to put together a verbal intervention model that fits your personality and with which you feel comfortable. The best preparation for potential crisis situations is to have spent some time thinking about your own intervention model. It is too late when you are in the middle of a crisis and wondering how to defuse the client who is acting out. The chapter provides you with a user-friendly model of verbal intervention.

References

Adler, R., Towne, N. (1993). *Looking out, looking in.* Orlando: Holt, Rinehart, and Winston.

Aguilera, D. (1994). *Crisis intervention theory and methodology* (7th ed). St. Louis: Mosby.

Coloroso, B. (1994). *Kids are worth it: Giving your child the gift of inner discipline.* Toronto: Summerville House.

Egan, G. (1994). *The skilled helper: A problem-management approach to helping.* Pacific Grove: Brooks/Cole.

France, K. (1990). *Crisis intervention a handbook of immediate person-to-person help.* Springfield, IL: Charles C Thomas.

Gossen, D. (1993). *Restitution: Restructuring school discipline facilitator's guide.* Chapel Hill, NC: New View Publications.

Howatt, W. (1996). *A teacher's survival guide for the 21st century: How to manage the classroom more effectively.* Athens, ON: Hindle.

Medler, A. (1992). Lecture notes, education seminar, McGill University.

Meichenbaum, D. (1977). *Cognitive-behavioral modification: An integrative approach.* New York: Plenum Press.

Scissons, E. H. (1993). *Counseling for results: Principles and practice of helping.* Pacific Grove, CA: Brooks/Cole.

Todd, J., Bohart, A. C. (1994). *Foundations of clinical and counseling psychology.* New York: HarperCollins.

Chapter 22

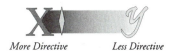

More Directive   Less Directive

Eclectic Suicide Intervention

From the first conversation with a suicidal individual, a counselor immediately takes an active role and some responsibility in preventing the client from choosing suicide.

Aguilera, 1994

MAIN IDEAS

In the field of counseling we often deal with people in extremely challenging situations. Unfortunately, you may find a client to be in what I call the "hopeless stage." What this means is that the client is thinking that life has no hope and finding it hard to face going on. This can lead to thoughts about ending one's life. We call this "suicide thinking."

The purpose of this chapter is to provide you with facts about suicide, cues to recognizing some of the warning signs, and a model for what to do if you have a suicidal client. I believe it is important for all new counselors, not to mention those with experience, to be aware of their limitations. This chapter has been written to help you stabilize a client only, so that you can proceed to get them appropriate help. Let me be clear, *this is only for intervention, not treatment.*

That having been said, you may, at some point, have to deal with a suicidal client. The following points provide some basic factual information about suicide and its impact on society.

- "When the desired assistance is not readily available, some individuals resort to life-threatening behavior as a dramatic 'cry for help.' Two-thirds of all suicide attempts are actually pleas for attention; they are intended to end in rescue rather than in death" (France, 1990).

- "Suicide is the eighth leading cause of death in America; its incidence is on the rise among all age groups, especially among youth, where it has become the second most frequent cause of death. More people die from suicide than homicide" (Maxmen and Ward, 1995).

- "Suicide lethality is the probability that the individual will kill himself or herself in the immediate future. Assessing this possibility is a mandatory activity with the suicidal client" (France, 1990).

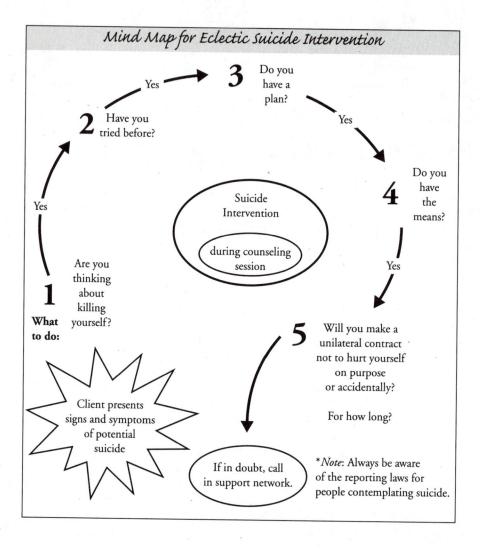

Mind Map for Eclectic Suicide Intervention

- "Ten to twenty percent of all suicide attempters eventually kill themselves" (Gabbard, 1994).

- "Approximately eighty percent of all individuals who kill themselves communicate their suicidal intent prior to their deaths" (France, 1990).

- "People who have already made a suicidal attempt are at higher risk, but only 2 to 4 percent kill themselves in the next five years after an attempt" (Maxmen and Ward, 1995).

- Maxmen and Ward (1995) indicate that a greater number of adults suffer from depression than adolescents. The incidence of suicidal thoughts by adults is also somewhat higher than in adolescents. But statistics also show that a higher percentage of adolescents with suicidal thoughts attempt suicide.

- "Higher demographic risks are male, over 40, alone, socially alienated, with chronic disease, obsessive-compulsive or perfectionist tendencies, and/or substance abusers" (Maxmen and Ward, 1995).

- France (1990) states the primary means of assessing suicide risk is the client's verbal communication. This may be either a direct statement ("I wish I were dead") or an indirect statement ("I'm a loser").

As a human services counselor who see a diversity of clients, it is imperative that you be able to recognize the warning signs of suicide and be prepared to intervene immediately. Any time you believe a client may be at risk for committing suicide, you must **refer** the individual to a professional (physician, psychologist, supervisor who specializes in treating this population).

The main considerations to remember when dealing with a suicidal client are: communication skills, symptoms and risk factors of suicidal clients, and degree of suicidal risk. In the event you ever need to do an intervention with a suicidal client, you need to be aware that the main tool you have is communication—the spoken word.

Communication Skills

Peach and Reddick (1991) recommend that the counselor who is talking to a client identified as a possible suicide risk can do two things: (1) *instill hope,* and (2) *reduce the client's anxiety.* The counselor uses an array of communication skills to achieve this, including listening, showing empathy, probing, and asking assessment questions. One of the best skills of communication is the ability to listen. Only through listening will you get the information you need—for example, client strengths and favored lifestyle—to address the suicidal thinking. You need this information if you are to have a chance of building rapport with the client.

Symptoms and Risk Factors

If you recognize in your client the following signs and symptoms that indicate the presence of risk for suicide, you will need to buy time for a professional referral. It is important to point out that the listed risk factors are not law; they are only guidelines.

- Aguilera (1994) explains that professionals should always monitor for depression. The majority of people who kill themselves are depressed.

- Maxmen and Ward (1995) also provide a list of symptoms potentially related to suicidal thinking: restlessness, sleep problems, change in eating habits, difficulty concentrating, loneliness, loss of interest in usual activities, fatigue, lack of interest in appearance, withdrawal from family or friends, feelings of hopelessness and helplessness.

- France (1990) provides the following list of risk factors:
  Substance abuse (higher suicidal risk)
  Family history of suicide
  Self-multilation (cuts, burns)
  Recent or impending loss (separation/divorce, death, rejection)

Loss of job or financial loss

Social isolation (no support systems and/or cultural differences)

Self-destructive behaviors (frequent accidents)

Preparation for death (making a will, giving things away), or acquiring the means (rope, gun)

- Aguilera (1994) states that "additional risk factors may occur when the client exhibits increased levels of tension, anxiety, guilt, shame, rage, anger, hostility, or revenge."

- Maxmen and Ward (1995) suggest that one of the most dangerous times is when the individual displays sudden, unexpected happiness and activity following prolonged depression. This may indicate that the person is relieved because the decision to commit suicide has finally been made.

- Other factors to take into consideration are some specific conditions associated with adolescent suicide. In 1986, Bratina suggested the following as adolescent risk factors: "families plagued by divorce; communication barriers between parents and adolescents; families with two careers; drug and alcohol addiction; pressures from parents, school, and friends; families who move a lot; fear of future jobs; and personal relationship problems" (Peach and Reddick, 1991). It should also be noted that suicides by adolescents may be more spontaneous than those by adults.

TECHNIQUES

Schniedman (1984) states that the goal of suicide prevention is to relieve the distress and pain so that the client is able to return to a "normal state." As a human services counselor, you need an intervention that will help to stabilize the situation until you are able to get professional assistance. For the purposes of this text, we have chosen to use a five-step suicide intervention model described by Wubbolding in his book *Reality Therapy Training Workbook* (1990), which includes a copy of his article "Professional Ethics: Handling Suicidal Threats in the Counseling Session" (refer to mind map for visual illustration). The model employs a series of questions to use in intervening.

Question 1   *"Are you thinking about killing yourself?"* The human services counselor asks this question of the client when it is suspected that the client is thinking about suicide or if the client has expressed it (Wubbolding, 1986). Gernsbacher (1984) states that clients will feel relieved that they can talk about it. The client may only be *thinking* about suicide, with no intention of going through with it. Still, the counselor needs to continue exploring the client's concerns and the circumstances that led to the suicidal ideation. Any information is important. The counselor also continues assessing the client.

Question 2   *"Have you tried before?"* The counselor asks if the client has attempted suicide in the past. Wubbolding (1986) states that past attempts are the best predictor of suicide. If the client has earlier attempted suicide, the individual is at high risk. The counselor continues using communication and exploration skills. For example, ask when, what happened, and what gave the client hope at that time.

Question 3 *"Do you have a plan?"* Gabbard (1994) states that suicidal clients should be asked outright if they have a plan. The counselor determines the client's plan and assesses the danger. It is necessary to ask what method the client is going to use. Examples:

- How will you do it? The more deadly the method, the more crucial it is to obtain medical assistance (France, 1990). The professional also explores if the client has a time, place, and location chosen.

- When are you planning to do it? Where? The danger is increased if the client's plans involve physical isolation, no matter what the method. Isolation decreases the potential for intervention (France, 1990).

Question 4 *"Do you have the means?"* If the client has access to the means, the risk is greatly increased. The counselor asks enough questions to find out if the client has a detailed, individualized plan (France, 1990). The client may say: "I have a gun, and tomorrow after everyone has gone to work I'm going to load the gun, walk into the woods, and shoot myself." There is further risk if the client has made final arrangements, such as writing or changing a will, giving away possessions, putting finances in order, or writing a note. Clients who do not have a specific plan may still be at risk, as they may act on impulse (especially youth).

Question 5 *"Will you promise not to kill yourself, on purpose or accidentally? For how long?"* France (1990) suggests that the ultimate goal of the helper is to encourage the client to make *a written contract for self* that leads to a no-suicide decision. The professional relies on personal question skills to achieve this end. If the client consents, it decreases the danger. A written contract is more powerful for the client than a verbal contract. Example:

I promise myself that I will not hurt myself, on purpose or accidentally, for the next _____ (period of time), until my counselor is able to assist me to get help with my situation.

Name: _____ Date: _____

If the client will not consent to a contract, the counselor must immediately break confidentiality and contact the appropriate professionals (a supervisor, the authorities). This follows the ethical guidelines of the Canadian Guidance and Counselling Association (Schulz, 1994, p. 48), which state:

When the counselee's condition indicates that there is *clear and imminent danger* [italics added] to the counselee or others, the member must take reasonable personal action or inform responsible authorities. The member should consult with other professionals and should only assume responsibility for the counselee's action after careful deliberation.

Summary

This chapter has been written for the purpose of intervention only. It is not a basis for treatment of clients who are suicidal. It is important for counselors to know their own limitations. I strongly recommend that, before starting employment, you discuss with your supervisor the agency's guidelines for dealing with potentially suicidal clients.

It is important that as a counselor you have sufficient skills to stabilize a suicidal client until you are able to obtain assistance. I recommend that you memorize the model provided by Wubbolding, so that in times of suicidal crisis you are ready to meet the challenge.

References

Aguilera, D. C. (1994). *Crisis intervention theory and methodology.* St. Louis: Mosby.

France, K. (1990). *Crisis intervention: A handbook of immediate person-to-person help.* Springfield, IL: Charles C Thomas.

Gabbard, G. O. (1994). *Psychodynamic psychiatry in clinical practice.* Washington, DC: American Psychiatric Press.

Maxmen, J. S., Ward, N. G. (1995). *Essential psychopathology and its treatment.* New York: Norton.

Peach, L., Reddick, T. L. (1991). Counselors can make a difference in preventing adolescent suicide. *The School Counselor 39*:107–109.

Schulz, W. E. (1994). *Counselling ethics casebook.* Ottawa: Canadian Guidance and Counselling Association.

Wubbolding, R. (1990). *Reality therapy training workbook.* Cincinnati: Center for Reality Therapy.

Chapter 23

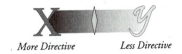

More Directive   Less Directive

Brief Counseling (Non-Crisis)

A small change can have far-reaching ripple effects.
de Shazer, 1985

HISTORY

Brief counseling is an action-oriented approach that is beneficial for clients with specific, concrete problems. Littrell (1990) teaches that brief counseling developed because of the demand for low-cost treatment and the emphasis towards a more socially oriented approach.

Brief counseling is based on the idea that the client has a ready supply of available resources. The purpose, then, is to assist clients to use their resources and to build on their past experiences of success to help deal with their present concerns (Littrell, 1990). Brief counseling is a pragmatic, structured, directive, time-limited approach that may consist of as little as one session or as many as six sessions.

MAIN IDEAS

Brief counseling involves leading the client to take action by choosing among available alternatives for resolving specific problems. The focus always remains on the client's strengths, resources, and goals. Brief counseling is not recommended for psychotic behavior, suicidal intentions, addictions, potentially violent individuals, or people with severe character disorders. People who may benefit from brief counseling are those suffering from anxiety, depression, or a situational problem.

TECHNIQUES

The brief counseling process presented here is based on the model delivered by Littrel (1990), which involves four steps that lead to change.

Step 1 *Define the problem.* The counselor uses various communication skills to gather information and define the client's specific problem. Skills used include open-ended questioning, paraphrasing, listening, and summarizing. Examples:

• What brings you here?

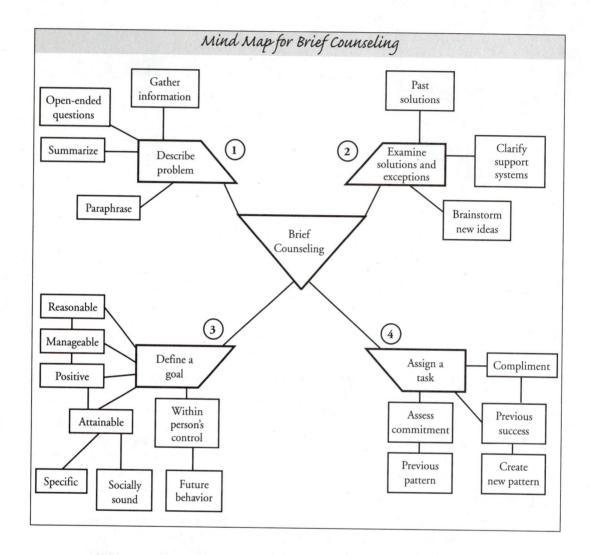

Mind Map for Brief Counseling

- What would be useful to help me to understand this situation?
- How do you do (a specific behavior, e.g., depression)?
- How does this create a problem for you?
- Describe to me in one sentence what you have told me.

Step 2 *Examine attempted solutions and exceptions.* In this step, the counselor explores past solutions used by the client in an attempt to solve the problem. The counselor also explores the client's support system and helps the client generate new ideas. Examples:

- When you have been faced with a similar situation in the past, what worked for you then?

- What else have you thought about trying?
- What have your family and friends suggested you try?

Step 3 *Define a goal.* The counselor helps the client set a goal. The goal needs to be reasonable, meaningful to the client, positive, specific, manageable, socially sound, and within the client's control. The counselor also instills motivation in the client by asking the client to imagine the desired outcome of the goal (future behavior). Examples:

- What would be a reasonable goal to set for yourself?
- Imagine you have resolved this situation (e.g., in three week's time). Describe how you are thinking, feeling, and acting.

Step 4 *Assign a task.* The counselor compliments the client on efforts made to solve the problem, which promotes acceptance of the intervention. This step also involves assessing the client's degree of commitment and then assigning tasks, depending upon the client's goals. Examples:

- If the client's goals are vague, the counselor assigns an observation task to clarify the client's goals ("To begin reaching your goal, I want you to find out what you really like about school").
- If clients have had earlier success, the counselor asks them to self-monitor their behavior ("To begin reaching your goal, pay attention to what you do when you overcome the urge to get depressed").
- If the client needs to create new patterns of behavior, the counselor suggests doing something different ("To begin reaching your goal, do something different, something fun. Remember, it cannot be illegal, immoral, or hurtful to yourself or others").

Summary

When you know that you will be seeing a client for only one to five sessions you may find that, with a client who is motivated to change, you can accomplish a significant amount using the brief therapy model in conjunction with your own counseling orientation.

In today's world, clients want quick results and will usually not be interested in waiting long periods of time. The client with the resources and the ability to be motivated can benefit a great deal from brief counseling. "If it ain't broke, don't fix it" is an old adage that describes what brief therapy stands for. Brief therapy allows a client to get back on track quickly.

References

deShazer, S. (1985). *Keys to solutions in brief therapy.* New York: Norton.

Littrell, J. (1990). Brief therapy. Paper presented at the conference for the Canadian Guidance and Counselling Association (Atlantic Chapter), Dartmouth, NS, June.

Chapter 24

More Directive   Less Directive

The Damage of Abuse

Abuse denotes an unequal power relationship.

Gilliland and James, 1996

This chapter deals with three categories of abuse: physical, sexual, and emotional. Nicarthy and Davidson (1989) provide the following definitions: *abuse* is any act that results in the mistreatment of another person. Abuse can happen to anyone, male or female. *Physical abuse* is characterized by a pattern of physically violent assaults. *Sexual abuse* is mistreatment in the form of unwanted sexual acts or demands. *Emotional abuse* involves belittling and eroding a person's self-esteem through verbal assaults, threats, and other acts.

PHYSICAL ABUSE: MAIN IDEAS

The common factor underlying all forms of abuse is the misuse of power and control. Physical abuse occurs when a more physically powerful or aggressive person takes advantage of a less powerful person through the perpetration of various physical acts. These acts may include slapping, hitting, shoving, punching, choking, using cigarettes to inflict burns to the body, throwing objects (such as furniture), or using hangers, belts, or other weapons to beat the victim. These incidents may lead to the ultimate act of violence—murder (Ontario Ministry of Community and Social Services, 1987; Walker, 1994).

PHYSICAL ABUSE: DEFINITIONS

The 1995 Canadian Criminal Code (section 265) states that a person commits an *assault* when:

- (a) without the consent of the other person, he applies force intentionally to that other person, directly or indirectly;
- (b) he attempts or threatens, by an act or gesture, to apply force to another person, if he has, or causes the other person to believe on reasonable grounds that he has, the present ability to effect his purpose; or

- (c) while openly wearing or carrying a weapon or an intimation thereof, he accosts or impedes another person. (Rodrigues, 1994)

Note: Most physical abuse is perpetrated by males, so that the use of the male pronoun is, unfortunately, accurate (Maguire and Radosh, 1999).

Physical abuse is defined by Hampton and colleagues (1993) as

any physical injury inflicted by other than accidental means that causes or creates a substantial risk of death, disfigurement, impairment of physical health, or the loss or impairment of the functioning of any bodily organ.

PHYSICAL ABUSE: WARNING SIGNS

Battering is defined as "the establishment of control and fear in a relationship through physical violence and other forms of abuse" (Cybergrrl, 1996). A battered woman is "one who has been physically, sexually and/or seriously (emotionally) abused by a person with whom she is in, or has been in, an intimate, romantic relationship" (Walker, 1994). Beagan (1992) estimated that at least 1 in 10 women involved in a domestic relationship is abused, and that 1 in 2 women will be battered at some time in her life. As many as 4000 American women are beaten to death each year. Most of these deaths occur when she ends, or tries to end, the relationship (Beagan, 1992). Physical violence within a relationship usually follows a typical pattern. Walker (1994) defines this pattern as the *cycle of violence* (Figure 24–1). Note that the figure represents the cycle but does not reflect intensity and duration, which Walker spends a great deal of time recording.

In the first stage of this cycle, tensions begin to build. The abuser experiences a decreasing ability to cope with various stressors and the relationship becomes

Figure 24–1.

The cycle of violence (adapted from Walker, 1994)

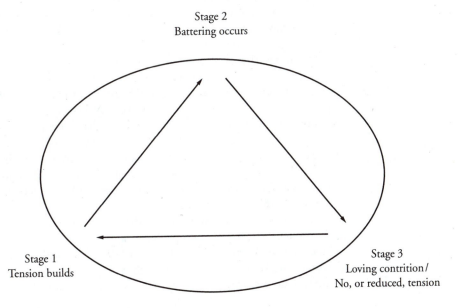

Stage 2
Battering occurs

Stage 1
Tension builds

Stage 3
Loving contrition /
No, or reduced, tension

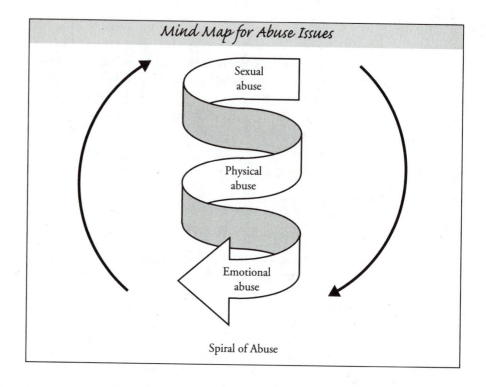

Mind Map for Abuse Issues

Sexual abuse

Physical abuse

Emotional abuse

Spiral of Abuse

strained. During the second stage, violence erupts. The third stage is marked by a period of profuse apologies and a return to calm. Abusers typically feel very sorry for their behavior and promise it will not occur again. This stage is often referred to as the "honeymoon phase." After a period of time, the relationship generally returns to the first stage and the cycle begins again. As the cycle continues, the violent episodes increase in severity and the period of time between violent episodes decreases (Walker, 1994; Gilliland and James, 1996).

As a human services counselor, it is important to be able to recognize the signs of physical abuse and be prepared to intervene. When clients approach a professional for help, it is imperative that response be immediate. *The counselor must believe, and nonjudgementally accept, the client.* Gilliland and James (1996) state that most women will only seek help after the situation has become exceedingly dangerous. A study by Statistics Canada (1994) found that 75% of victims do not report abuse until at least nine violent episodes have occurred. Other sources find that it may take an average of thirty-five episodes before help is sought (Martin and Moore, 1995).

Many male batterers suffer from low self-esteem and have inadequate coping skills. Rather than dealing with problems appropriately, they vent their frustrations on their female partners (Thorne-Finch, 1992). Thorne-Finch cited a study by Goldstein and Rosenbaum that found abusive men are more likely to perceive their partner's actions as threatening to their self-image.

Not all perpetrators of physical abuse are male; occasionally men are physically abused by their wives or significant others. Men may simply be more reluctant to report abusive incidents because of the way society views masculinity. Because women are generally physically weaker than men, it is unusual for a woman to control a man by force (Nicarthy and Davidson, 1989).

Warning signs of battering may include various physical indicators (Ontario Ministry of Community and Social Services, 1987):

- Unexplained bruises, welts, or cuts
- Large number of scars
- Burn marks that appear to have been caused intentionally
- Unexplained fractures or a number of old, healed fractures
- Attempts to hide wounds
- Large number of "accidents" at home

Other sources add to the above list of indicators:

- Depression and low self-esteem (Martin and Moore, 1995)
- Suicidal thoughts and gestures (Martin and Moore, 1995)
- Symptoms of posttraumatic stress disorder (PTSD) (Gilliland and James, 1996)
- Fear for children's safety (Gilliland and James, 1996)
- Adherence to strict sex roles (Gilliland and James, 1996)
- Social isolation (Gilliland and James, 1996)
- Self-blame, humiliation (Gilliland and James, 1996)
- Family history of abuse (Gilliland and James, 1996)
- Wide range of chronic physical complaints (Ontario Medical Association [OMA], 1991)
- Substance abuse (OMA, 1991; Gilliland and James, 1996)

Walker (1994) added the following personal and psychological factors:

- Fear of retribution
- Denial
- Repression
- Dissociation
- Self-blame
- Shame, guilt, and lack of trust
- Cognitive confusion and/or dysfunction

SEXUAL ABUSE: MAIN IDEAS

Sexual abuse is one of the most serious and tragic crises counselors may encounter. Sexual abuse carries with it enormous social and moral prejudices; therefore, it is important to note that crises resulting from sexual abuse are different in nature, intensity, and extent from other forms of crisis (Pearson, 1994).

It is very difficult to predict how a survivor of sexual abuse will react. Some survivors appear to take sexual abuse in stride, whereas another person subjected to a similar experience virtually falls apart (Beagan, 1992). As with physical abuse, it is well known that men are the main perpetrators of all forms of sexual abuse. Fewer than 2 percent of sexual offenders are female (DeKeseredy and Hinch, 1991). Of the sexual assaults committed, only about 20 percent are perpetrated by strangers. The remainder of assaults are committed by individuals known to the victim (AMA, 1995). According to Gilliland and James (1996), the rate of sexual abuse has increased with the development of widespread divorce and remarriage, as well as the increasing prevalence of short-term commonlaw relationships (which increases the probability that individuals will be exposed to sexually violent partners). It is estimated that 1 in 3 women, 1 in 9 men, and 1 in 4 children in the United States have been sexually assaulted (Sexual Assault Crisis Center, 1995). A study by Statistics Canada (1993) shows that at least 39% of women have been sexually assaulted one or more times in their lives.

The vast majority of sexual assaults involve power relationships between males and females. It is estimated that 1 in 8 wives is the victim of marital rape, and the incidence of rape within a dating relationship is even higher (Thorne-Finch, 1992). Contemporary sociocultural dynamics produce some men who feel such an absence of power in their lives that they develop a need to take it. These men come to believe it is their right to take it; they rationalize and justify their behavior even though they have invaded another person's life and body. The rapist's justification for committing the ultimate act of humiliation and degradation becomes forcing the person to submit (Gilliland and James, 1996). Rape is a traumatic event to which no one would consent. Our job is not to judge the rape or its impact, but to help our clients move on.

Sexual assault is a social phenomenon. They identify four causes: (1) gender inequality, (2) pornography, (3) social disorganization, and (4) legitimization of violence. Our cultural advances have lagged behind our technological and social evolution. Attitudes about sex, sexuality, and sexual assault have been slow to change.

Four factors that affect the speed of recovery for a sexually abused individual include: (1) the degree of intimacy between the survivor and the assailant, (2) the age of the individual at the onset of abuse and the extent of time over which the abuse occurred, (3) the intrusiveness of the abuse, and (4) the manner in which the individual was inducted into the sexual activity (Hampton et al., 1993).

SEXUAL ABUSE: DEFINITIONS

When confronted by a client who has been abused, it is important to have clarity about professional and legal definitions. *Sexual abuse* has been defined as

> any unwanted sexual contact or attention achieved by force, threats, bribes, manipulation, tricks, coercion, or violence. Sexual assaults may be physical or non-physical

acts which include, but are not limited to, unwanted touching of an intimate part of another person, such as a sexual organ, buttocks or breast, and bodily penetration by a foreign object. (SACC, 1995)

Rape is defined as

a sexual invasion of the body by force, an intrusion into private, personal, inner space without consent—in short, an internal assault from one of several avenues and by one of several methods [that] constitutes a deliberate violation of emotional, physical, and rational integrity and is a hostile, degrading act of violence. (Brown-miller, 1985)

SEXUAL ABUSE: WARNING SIGNS

Personal and psychological factors that tend to affect the survivor's responses to sexual assault, and to the recovery process, include the following. People who are assaulted:

- Fear for their life. Many victims do not react to the sexual aspects of the crime, but rather to the terror and fear involved (SACC, 1995).
- May respond by exhibiting no emotions—appearing unaffected or dissociated (AMA, 1995).
- Feel humiliated, demeaned, degraded, and helpless (AMA, 1995)
- May experience symptoms of posttraumatic stress disorder (PTSD) (AMA, 1995)
- Depression (AMA, 1995)
- Fear of pregnancy and sexually transmitted diseases (STDs) (AMA, 1995)
- May suffer immediate physical and psychological injury, as well as long-term trauma (Gilliland and James, 1996)
- May experience impaired sexual functioning (Gilliland and James, 1996)
- Male victims of sexual abuse often question their sexuality (AMA, 1995)
- May blame themselves and feel guilty (AMA, 1995)
- May experience difficulty relating to and trusting others; especially men (AMA, 1995)
- May feel intense anger or hatred toward the assailant. This may be exhibited through increased levels of aggression (AMA, 1995)
- Will never be the same, even though most survivors, over time, develop ways to recover, cope, and go on with their lives (AMA, 1995)
- May be fearful of going to the police or a rape crisis center (AMA, 1995)
- May be reluctant to discuss the assault with members of the family, friends, and others, because of the risk of rejection and embarrassment (AMA, 1995)
- May have suicidal tendencies (Pearson, 1994)

EMOTIONAL ABUSE: MAIN IDEAS AND DEFINITION

Engel (1990) defined emotional abuse as

> any kind of abuse that is emotional rather than physical in nature. It can include anything from verbal abuse and constant criticism to more subtle tactics, such as intimidation, manipulation and refusal to ever be pleased. Emotional abuse is like brainwashing in that it systematically wears away at the victims self-confidence, sense of self-worth, trust in his/her perceptions and self-concept.

Nicarthy and Davidson (1989) state that the best way to control a person is through the mind and the best way to achieve this is through brainwashing. Emotional abuse cuts to the very core of a person, creating scars that may be far deeper and more lasting than physical ones.

The subtlety with which emotional abuse can be delivered encourages society to minimize its effects; yet the experience of emotional abuse is considered to be "more painful and damaging than physical abuse" (Hampton et al., 1993). It is considered to be the common theme underlying all abuse, including physical and sexual abuse. It would appear that physically abusive men employ a variety of emotionally abusive strategies in relationships. Folingstad and colleagues (1990) studied 234 women with a history of physical abuse. Ninety-nine percent of these women reported experiencing at least one form of emotional abuse at some point in their marriage. Of the types of abuse reported, 90% reported experiencing ridicule or verbal harassment, 79% experienced restriction or isolation of activities, 74% experienced threats of abuse, 73% experienced severe jealousy or possessiveness, 59% experienced property damage, and 48% experienced threats to change or end the marriage. Restriction/isolation, jealousy/possessiveness, and ridicule/harassment were the most commonly endured, with 60% of the women reporting experiencing these forms of abuse at least once a week.

Thus, it would appear physical abuse and emotional abuse go hand in hand. However, emotional abuse can and does appear in isolation. In fact, some form of emotional abuse is said to be present in the vast majority of adult-adult and adult-child relationships (Hampton et al., 1993). The fact is, emotional abuse can be so engrained in the language that it becomes normalized and accepted. Much of the abuse I see is emotional rather than the more commonly acknowledged sexual or physical abuse. Emotional abuse is a problem that we in this field need to recognize and address.

Emotional abuse operates by instilling fear, increasing dependency and/or damaging self-esteem (Hampton et al., 1993). Emotional abuse includes the following behaviors:

- *Domination*— the need to be in charge and to control another person's every action. This person needs to get his/her own way and will often resort to threats to achieve this (Engel, 1990).
- *Verbal assaults*—berating, belittling, criticizing, name calling, screaming, threatening, blaming, and using sarcasm and humiliation. Verbal abuse as-

saults the mind and the spirit, and is extremely damaging to the victim's self-esteem and self-image (Engel, 1990).

- *Abusive expectations*—involves unreasonable demands being placed on the victim with the expectation that the victim will do everything necessary to satisfy the abuser's needs. These abusers demonstrate a constant need for undivided attention, demand frequent sex, or expect the victim to spend all free time with them (Engel, 1990).

- *Emotional blackmail*—involves manipulating another person into doing what the abuser wants by playing on the victim's fear, guilt, or compassion (Engel, 1990). This may involve withdrawing affection, ignoring, and/or sulking (Thorne-Finch, 1992).

- *Sexual harassment*—involves any unwelcome sexual advances or physical or verbal conduct of a sexual nature. Any situation in which a person is pressured into being sexual against the will is defined as sexual harassment (Engel, 1990). This may include promises of rewards for compliance or threats for non-compliance (Thorne-Finch, 1992).

- *Threats*—to harm self, partner, friends, relatives, pets, etc. (Hampton et al., 1993; Walker, 1994).

- *Jealousy and possessiveness*—accusations or recriminations of infidelity (Walker, 1994)

- *Minimization and denial*—involves efforts to downplay the extent or impact of violence or abuse, often by questioning the partner's sanity, perceptions, or feelings (Walker, 1994).

- *Pornography*—portrays individuals (usually women and children) as sex objects. Often implies that individuals portrayed enjoy violence in conjunction with sexually explicit activities (Thorne-Finch, 1992).

- *Economic deprivation*—involves attempts to control the finances of, or increase the financial dependency of, an individual (Hampton et al., 1993).

- *Neglect*—a condition in which an individual fails to provide the essentials necessary for the physical, emotional, and intellectual development of another person(s). It may also involve a failure to provide a safe, non-injurious environment or necessary medical care (Hampton et al., 1993).

Another form of emotional abuse is emotional incest. Emotional incest is typically found only in parent-child relationships. Love and Robinson (1990) define *emotional incest* as:

A style of parenting in which parents turn to their children, not to their partners, for emotional support. Children are forced to suppress their own needs to satisfy the needs of their parents. This role reversal leads to a lack of adequate protection, guidance and discipline.

There are two ways parent-child relationships can be out of line (Love and Robinson, 1990): (1) they can be *estranged,* with too much distance between parent and child; and (2) they can be *enmeshed,* with too close a relationship between parent and child. Enmeshed relationships are quite common, with close relationships between a child and the parent of the opposite sex being so widespread that popular terms have been coined for them, such as "Daddy's little girl" and "Momma's boy." Emotional incest occurs when there is an exceptionally high degree of enmeshment between parent and child. Families in which emotional incest occurs can be identified by the following characteristics (Love and Robinson, 1990):

- The parent lacks sufficient companionship and support.
- In a two-parent household, the parent who is less involved with the child feels resentful.
- In families with two or more children, there is more than the normal amount of sibling rivalry.

There are three types of emotionally incestuous parental roles. They are:

1. *Neglectful parent*—who does not meet the basic needs of the child. This may occur as a result of alcohol or drug addiction, or workaholism. In this type of relationship, the child is forced to take on many parental responsibilities. As a result, the child becomes self-reliant, invulnerable, and carries buried anger and resentment.

2. *Critical/abusive parent*—who is hypercritical of the child. In the case of two-parent families, the enmeshed parent is extremely permissive; the other parent becomes abusive, due to jealousy. In this type of relationship, the child becomes both surrogate spouse and scapegoat.

3. *Sexualizing parent*—this type of relationship is referred to as the "Cinderella syndrome." The child's relationship with the bonded parent is romantic, similar in many ways to a dating relationship. The bonded parent is sexually and emotionally fixated on the child. This is a much more intense and damaging relationship. In some cases, sexualizing parents indulge in sexual relationships with their children; others do not.

EMOTIONAL ABUSE: EFFECTS

The effects of emotional abuse and emotional incest are many and varied, and they affect every aspect of an individual's life. For emotional incest, they include (Love and Robinson, 1990):

- guilt
- chronic low-level anxiety
- social isolation
- self-image problems (including fear of rejection, feelings of inferiority, perfectionism, and the compulsive need to succeed)
- identity problems (including diffuse sense of identity, inability to separate from parent, and personal boundary problems)

- relationship problems (including fear of commitment, lack of romantic attraction, conflicts between spouse and parent, attraction to self-centered partner)
- sexual problems
- Emotional abuse (Engel, 1990) (including lack of motivation, confusion, difficulty making decisions or concentrating, low self-esteem, feelings of failure, worthlessness, and hopelessness, and self-blame and self-destructiveness)

INTERVENTION TECHNIQUES

Before a human services counselor intervenes, it is important to review with the client the limits of confidentiality and the obligation to report incidents to the proper authorities. All disclosures of abuse of a minor between the ages of zero and sixteen must generally be reported within 24 hours of being informed. It is recommended that the police also be notified of these incidents. As for disclosures of abuse against adults, we recommend that you report the incident to your supervisor, as well as keep extensive written reports of all information disclosed. Walker (1994) identifies two goals when counseling abuse survivors: (1) help them to achieve violence-free living; and (2) help re-empower them to become survivors, not victims. When intervening, the counselor can use the following techniques.

1. *Active listening.* Listen to the client in an empathic and accepting manner. It is important that an abused individual feel that a safe, supportive environment is available. Accept the account of abuse calmly and nonjudgementally (Gilliland and James, 1996; Cybergrrl, 1996). Probably the two most important and practical things to an abused person are: (1) They are not to blame for the abuse; (2) you are willing to listen (SACC, 1995).

2. *Ensure safety.* Determine the lethality of the situation. Encourage and help the client to develop a safety or escape plan and commit to using it (Gilliland and James, 1996; Cybergrrl, 1996; Walker, 1994). Help clients to heighten their sensitivity to their own physiological danger warning signs and to heed these warnings (MacLeod, 1980; Walker, 1994). Example:
   - Have an extra set of clothes, pocket money, and important documents at a friend's house in case you need to leave quickly. It is important that you store only items the abuser won't notice to be missing.

3. *Educate and share information.* Teach the cycle of violence (see Battering). Discuss the dynamics of violence, and how abuse is based on power and control. Provide information about available services that help women and children, such as social services, women's shelters, counseling services, and legal advice and protection (Cybergrrl, 1996). It is important to inform the client what to expect when dealing with any of these services (e.g., explain what will happen when they decide to file charges). Make referrals to service agencies when applicable.

4. *Normalize feelings.* Tell clients that they are not alone, that there are others in the same type of situation. Acknowledge their fear about discussing the abuse.

Reassure them that nothing they do or say makes it OK for them to be hit, beaten, verbally berated, or sexually violated. They are not to blame. Reinforce them for seeking help. It takes strength to trust someone enough to talk about the abuse (Cybergrrl, 1996; SACC, 1995; Gilliland and James, 1996).

5. *Self-defense.* Learning these skills can help clients save their own life, or simply enable them to escape a dangerous situation. The client is usually taught simple procedures to better protect the self from physical or sexual assaults (Pearson, 1993). Help the client become aware of the local self-defense courses available (consult Yellow Pages, directories, local YMCA, etc.).

6. *Assertiveness training.* Learning these skills can increase clients' self-esteem and sense of control over their lives (Pearson, 1993). This involves learning that everyone has self-worth and personal rights that should not be infringed upon by anyone. It also involves teaching individuals to ask for what they want and to say no when they do not want to comply. Finally, it involves finding a balance between being passive and being aggressive (Bourne, 1990). (See Chapter 13 for more information on this.)

7. *Support.* Encourage clients to express their emotions. Abused individuals often harbor feelings of hurt, guilt, fear, and anger about the abuse. Realize they may experience grief with respect to the loss of the relationship, role, job, and/or lifestyle (Russell and Uhlemann, 1994). (See Chapter 15.) Allow them to make their own decisions, even if they are not ready to leave the abusive situation. The counselor will often need to exhibit high levels of patience and unconditional positive regard (see Chapter 3) when working with abuse victims. At many stages of the therapeutic relationship, the client may express dependency, ambivalence, and/or depression. To counteract this, the counselor should strongly reinforce the victim's attempts at decisionmaking, self-control, and expressions of personal power. The counselor should also be careful not to criticize the abuser, as the client will often have ambivalent feelings toward that person, and may resent a counselor who negatively characterizes the abuser. The counselor should cycle between asking open-ended questions and reflecting and clarifying the victim's feelings. The counselor should also be prepared to help the clients to confront their faulty and illogical perceptions of the abusive situation (Gilliland and James, 1996).

8. *Medical attention.* If clients have been physically or sexually abused recently, they may require medical attention. Many abuse victims will not seek medical attention on their own, and therefore may not receive desperately needed medical treatment. Seeking medical attention is equally important for documentation, if the client wishes to file charges against the abuser. For those who are sexually abused, it is also important to preserve evidence and to be examined for possible pregnancy and STDs (SACC, 1995).

9. *Referral.* Human services counselors must know their own limitations. If the client's trauma is outside you area of training and expertise, it is your ethical responsibility to refer the client to the appropriate resource.

Summary

This chapter identified the main points to be aware of when dealing with abused clients. Remember that all three types of abuse may occur in combination, or on their own. You must be able to recognize the warning signs of each and be able to intervene quickly and tactfully. The counselor should help the client recognize and label the abuse, and then develop a plan to deal with it. Gilliland and James (1996): "The residual trauma affecting individuals depends on many variables, including, but not restricted to: the type of abuse experienced, time frame and intensity of abuse, attitudes of family and support of persons following disclosure/discovery, and effects of legal and police proceedings surrounding disclosure/discovery."

Finally, it is important to be active in educating others about what constitutes abuse and to develop a *no tolerance policy.* Professionals are required to be fully aware of the laws regarding reporting of abuse. For example, in Nova Scotia the law states that any adult who hears (firsthand or secondhand) that a child under the age of 16 (or 18, if protected by the courts due to early abuse history or threats), has a *legal* responsibility to report this abuse. Anyone not reporting this information can be charged under Nova Scotia law. It is imperative that you be fully informed of your own laws regarding abuse.

In closing, abuse is everyone's concern, and as human services counselors we are responsible for helping to support and protect our communities. No one has the right to knowingly hurt another human being. We all have the right to feel safe!

References

American Medical Association (AMA) (1995). Sexual assault in America. Available: http://www.ama-assn.org/public/releases/assault/action.htm.

Barnard, S. (1992). *The Barnard/Columbia women's handbook 1992.* New York: Columbia University Press.

Beagan, B. L. (1992). *Making changes: A book for women in abusive relationships.* Halifax: Nova Scotia Advisory Council on the Status of Women.

Bourne, E. J. (1990). *The anxiety and phobia workbook.* Oakland, CA: New Harbinger Books.

Brownmiller, S. (1975). *Against our will: Men, women, and rape.* New York: Simon & Schuster.

Cybergrrl Internet Media. (1996). Domestic violence: The facts. Available http://www.cybergrrl.com/dv/book/toc.html#toc

DeKeseredy, W. S., Hinch, R. (1991). *Women abuse: Sociological perspective.* Toronto: Thomson.

Engel, B. (1990). *The emotionally abused woman: Overcoming destructive patterns and reclaiming yourself.* New York: Fawcett Columbine.

Gilliland, B. E., James, R. K. (1996). *Crisis intervention strategies* (2nd ed.). Pacific Grove, CA: Brooks/Cole.

Hampton, R. L., Gullotta, T., Adams, G., Potter, E., Weisberg, R. (1993). *Family violence: Prevention and treatment.* Newbury Park, CA: Sage.

Love, P., Robinson, J. (1991). *The emotional incest syndrome: What to do when a parent's love rules your life.* Toronto: Bantam.

MacLeod, L. (1980). *Wife battering in Canada: The vicious cycle.* Ottawa: Canadian Advisory Council on the Status of Women.

Martin, D. G., Moore, A. D. (1995). *First steps in the art of intervention.* Pacific Grove, CA: Brooks/Cole.

Nicarthy, G., Davidson, S. (1989). *An easy to read handbook for abused women.* Seattle: Seal Press.

Ontario Medical Association (OMA), Committee on Wife Assault (1991). *Reports on wife assault.* Ontario: Author.

Ontario Ministry of Community and Social Services. (1987). *Child abuse prevention.* Toronto: Queen's Printer for Ontario.

Pearson, Q. M. (1994). Treatment techniques for adult survivors of childhood sexual abuse. *Journal of Counseling and Development 73:* 32–37.

Rodrigues, G. P. (Ed.). (1994). *Pocket criminal code: 1995.* Toronto: Carswell Thomson.

Russell, B., Uhlemann, M. R. (1994). Women surviving an abusive relationship: Grief and the process of change. *Journal of Counseling and Development 72:* 362–66.

Sexual Assault Crisis Center (SACC) of Knoxville, TN (1995). *What is sexual assault?* Available: http://www.cs.utk.edu/~bartley/sacc/what is SA.html.

Statistics Canada (1994). *Family violence in Canada.* Ottawa: Canadian Centre for Justice Statistics.

Thorne-Finch, R. (1992). *Ending the violence: The origins and treatment of male violence against women.* Toronto: University of Toronto Press.

University of British Columbia (1995). *Sexual assault is a crime.* Available: http://unixg.ubc.ca:880/lpalmer/women/wsosafe.htm.

Walker, L. (1994). The abused woman: A survivor therapy approach. In Walker, L., Lurie, M. (producers), and Schein, L. (director). *The Newbridge assessment and treatment of psychological disorders series.* [videotape]. New York: Newbridge Communications.

Chapter 25

More Directive   Less Directive

Addictions Counseling

God grant me the serenity to accept the things I cannot change, the courage to change the things I can, and wisdom to know the difference.

St. Francis

HISTORY

The history of humanity makes clear that we have always been in the process of discovering and using nature's gifts to enhance our physical and mental capabilities. Payne, Hahn, and Pinger (1991) state that, as *Homo sapiens* evolved, so too did our drug behaviors. They note archaeological evidence of drug use that dates back nearly 5000 years. Through time, humans have extracted opium from the poppy plant, cocaine from the leaves of the coca bush, and cannabis from the hemp plant. These substances were initially used for medicinal purposes, but contributed to concern about substance abuse when individuals began using them as a way of escaping from their personal problems.

With the eighteenth and nineteenth centuries came the industrial revolution and great technological advances. According to Prashant, a great variety of potent and harmful drugs, including narcotic substances, flooded the market. Machines that produced increasingly sophisticated goods, rapid developments in technology, and two world wars all contributed new opportunities to experience the effects of mood-altering substances as well as to produce drugs of varying potencies and effects.

The word *drug* refers to any chemical substance, aside from food, that upon entering the body changes the way the mind and/or body functions. In most industrial societies today, there is no clear distinction between legitimate drug use, misuse, and abuse. However, the misuse of drugs often occurs when an individual takes a prescription drug or an over-the-counter drug for reasons other than originally intended, or at a dosage other than recommended (Jacobs and Fehr, 1987).

MAIN IDEAS

It is likely all human services counselors, at some point in their career, will be faced with a client who has concerns with drugs or alcohol. This section provides a brief introduction to the field of addictions. I recommend that any person planning to

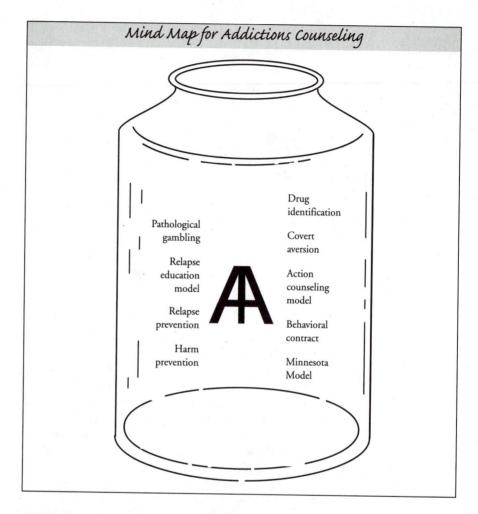

Mind Map for Addictions Counseling

Pathological gambling

Relapse education model

Relapse prevention

Harm prevention

Drug identification

Covert aversion

Action counseling model

Behavioral contract

Minnesota Model

AA

work in the field of addictions do further research and receive specialized training. In the field of addiction, professionals rely on the *Diagnostic and Statistical Manual of Mental Disorders* (1994) to determine identifying criteria. There are many types of drug addiction. However, the two main criteria used to determine if the client is dependent upon or abusing a drug (DSM-IV) are as follows.

Criteria for Substance Dependence

A. A maladaptive pattern of substance use leading to clinically significant impairment or distress, as manifested by three (or more) of the following, occuring within a 12-month period:

(1) tolerance, as defined by either of the following:
   (a) a need for markedly increased amounts of the substance to achieve intoxication or desired effect
   (b) markedly diminished effect with continued use of the same amount of the substance

(2) withdrawal, as manifested by either of the following:
- (a) the characteristic withdrawal syndrome for the substance (refer to Criteria A and B of the criteria sets for Withdrawal from the specific substances)
- (b) the same (or a closely related) substance is taken to relieve or avoid withdrawal symptoms

(3) the substance is often taken in larger amounts or over a longer period than was intended

(4) there is a persistent desire or unsuccessful efforts to cut down or control substance use

(5) a great deal of time is spent in activities necessary to obtain the substance (e.g., visiting multiple doctors or driving long distances), use the substance (e.g., chain-smoking), or recover from its effects

(6) important social, occupational, or recreational activites are given up or reduced because of substance use

(7) the substance use in continued despite knowledge of having a persistent or recurrent physical aor psychological problem that is likely to have been caused or exacerbated by the substance (e.g., current cocaine use despite recognition of cocaine-ionduced depression, or continued drinking despite recognition that an ulcer was made worse by alcohol consumption)

Specify if:

With Physiological Dependence: evidence of tolerance or withdrawal (i.e., either Item 1 or 2 is present)

Without Physiological Dependence: no evidence of tolerance or withdrawal (i.e., neither Item 1 nor 2 is present)

Criteria for Substance Abuse

A. A maladaptive pattern of substance use leading to clinically significant impairment or distress, as manifested by one (or more) of the following, occuring within a 12-month period:

(1) recurrent substance use resulting in a failure to fulfill major role obligations at work, school, or home (e.g., repeated absences or poor work performance related to substance use; substance-related absences, suspensions, or expulsions from school; neglect of children or household)

(2) recurrent substance use in situations in which it is physically hazardous (e.g., driving an automobile or operating a machine when impaired by substance use)

(3) recurrent substance-related legal problems (e.g., arrests for substance-related disorderly conduct)

(4) continued substance use despite having persistent or recurrent effects of the substance (e.g., arguments with spouse about consequences of intoxication, physical flights)

B. The symptoms have never met the criteria for Substance Dependence for this class of substance.

Dependency on a chemical is marked by the development of tolerance and withdrawal (Schuckit, 1989). Doweiko (1996) explains that tolerance to a drug will develop over time, as the body struggles to maintain normal functioning in spite of the presence of one or more foreign substances. Most drugs, if used long enough over a period of time, will bring about characteristics of withdrawal, and certain groups of drugs will produce physical symptoms when the person discontinues their use (Doweiko, 1996).

It is important to point out that *denial* is usually present with the majority of individuals who are experiencing problems with substance abuse. Evans and Sullivan (1995) state "the disease of addiction is a disease of denial." Much of the client's denial is based in what cognitive-behavioral therapy terms *dysfunctional thinking* (see Chapters 16 and 17). When clients can dispute their irrational thinking, they are in a position to address the psychological component of their addiction. It is important for the counselor to recognize that anyone who is chemically dependent is usually both psychologically and physically addicted. Psychological dependence is mentioned first because denial (a key component of psychological dependence) can exist at both conscious and unconscious levels, which makes psychological dependency very complex and challenging. A review of the literature reveals that denial is one of the ingredients that puts the client at risk for a potential relapse by providing the client with a false sense of security.

THEORETICAL MODELS OF ADDICTION

Historically, alcohol and/or drug dependency has been viewed as either a sin or a disease (Thombs, 1994). However in recent years, alcohol and drug abuse have also been seen as a pattern of maladaptive behaviors. Over the years there have been a number of theories and models developed in an attempt to explain the underlying reasons for addiction. Broadly speaking, they fall into the moral and medical categories.

In the past, addicts were seen as people who were morally unfit or lacking in willpower; their cravings for alcohol or drugs were looked upon as figments of their imagination. It was believed that they could overcome their problems if they could just pull themselves together (Gold, 1991). Excessive drinking or drug use was considered to be freely chosen behavior that created suffering for users and their significant others. Addicts were blamed for their undesirable behaviors and considered responsible for their abuse problems. In this model, punishment was relied upon to rectify past misdeeds and to prevent further chemical use (Thombs, 1994). It was not believed that the addict deserved care or help.

The American Medical Association classified alcoholism as a disease in 1957 (Doweiko, 1996). The disease model was first conceptualized by E. M. Jellinek (1960) as a way to understand alcoholism, and it has since been applied to the broader field of addictions. The work of Jellinek has had a profound impact on the evolution of the medical model of addiction. Jellinek suggests that addiction to alcohol progresses through four stages: (1) pre-alcoholic phase, (2) prodromal phase, (3) crucial phase, and (4) chronic phase.

Jellinek's Alcoholism Model

Phase I *Pre-alcoholic.* This phase is marked by the individual's use of alcohol for relief from social tension. Continued drinking for this purpose leads to a gradual
increase in physiological tolerance. The individual must drink larger quantities
more frequently to achieve the desired state.

Phase II *Prodromal.* This phase is marked by the development of memory blackouts, secretive drinking, a preoccupation with alcohol use, and guilt over drinking behavior. This phase also results in the increased use of defense mechanisms
such as denial.

Phase III *Crucial.* In this phase, physical addiction occurs. Other symptoms of the
crucial phase include loss of self-esteem, loss of control over one's drinking, social withdrawal in favor of alcohol use, self-pity, and neglect of proper nutrition.
The individual attempts to control drinking by entering into short periods of
abstinence, only to return to the use of alcohol. The individual becomes willing
to risk everything to continue drinking.

Phase IV *Chronic.* The symptoms of this phase include deterioration in moral behavior, drinking with inferiors, the development of motor tremors, an obsession with drinking, and, for some, the use of substitutes when alcohol is not
available.

Jellinek's theory was based on questionnaire information obtained from a sample of Alcoholics Anonymous members. Of the 158 questionnaires sent out, ninety-
eight (62 percent) were used to formulate his conclusions (Pratsinak and Alexander,
1992). Jellinek's model of alcoholism was originally developed as a theory of alcohol
abuse; however, it has been applied to virtually every other pattern of abuse.

The disease model of addiction accepts that addicts have a virtually irresistible
physical craving for the substance, and experience a loss of control over substance use
(Lewis, Dana, and Blevins, 1994). Addiction is seen as a progressive, irreversible disease with a biological component (Doweiko, 1996). In this view, substance abusers
(whether presently using drugs or not) are living with a disease, and any return to
use will immediately reactivate the disease process and substance dependency (Lewis,
Dana, and Blevins, 1994).

In 1960, Jellinek classified patterns of addictive drinking. He identified five sub-
forms of alcoholism using the first five letters of the Greek alphabet: alpha, gamma,
beta, delta, and epsilon.

Jellinek's five classifications

Alpha alcoholics are psychologically dependent on alcohol; however, they do not experience any physical complications. According to Jellinek, alpha drinkers can ab-
stain from alcohol use for short periods of time when necessary. They are relatively
stable, rarely progressing to a more serious form of alcohol use. If an alpha drinking
pattern were to escalate into another form of alcoholism, Jellinek believed it would
evolve into the gamma form of alcoholism.

Beta alcoholics experience psychological dependence as well as physical complications. They may demonstrate medical symptoms of chronic alcohol use, such as cirrhosis of the liver or gastritis. If beta alcoholics progress, they would be classified as gamma alcoholics.

Delta alcoholics demonstrate physical dependence on alcohol, including tolerance and cravings. However, delta alcoholics show few or no physical complications that could be traced to alcohol use.

Gamma alcoholics experience physical withdrawal symptoms, cravings, physical dependence, and develop various medical complications caused by alcohol use. Also, the gamma alcoholic demonstrates a progressive loss of control over alcohol use.

Epsilon alcoholics may best be described as binge drinkers.

Other Theories and Models

Inheritability or Genetic Theory

According to Gold (1991), any disorder that can be passed from one generation to the next has a biological basis. Research confirms that having a family history of alcohol abuse increases the risk of developing alcoholism by as much as 50 percent (Thombs, 1994). Similar patterns are seen among families in which a member abuses substances other than alcohol, such as cocaine or narcotics (Gold, 1991).

Cloninger, Gohman, and Sigvardsson (1981) conducted one of the most extensive research studies exploring the genetics of alcoholism by using the adoption records of some 300 individuals from Sweden. The researchers found that children of alcoholic parents were likely to grow up to be alcoholics themselves, even in cases where the children were reared by non-alcoholic parents almost from birth. Gold's (1991) research on families clearly shows that genetic factors play a role in the susceptibility to alcohol abuse behavior in certain individuals regardless of environmental or other contributing factors.

Set point theory

The body has a number of thermostats that keep things operating smoothly and work to protect the body (Gold, 1991). For example, if your weight drops below a certain level, the weight regulator in your brain will kick in, shutting down some systems and accelerating others in an effort to get you to eat more (Gold, 1991). This concept is known as *set point theory*. According to Gold (1991), individuals have different set points that determine their vulnerability to drugs or alcohol. Everybody's physiology is different. Some people break down toxic substances better or more efficiently than others. If you have a low set point for cocaine metabolism, then even a small dose might overwhelm your body's defenses, invade delicate tissue, and damage brain cells (Gold, 1991). For example, small amounts of cocaine can impel people with a low set point on the path to addiction.

Social learning theory

Social learning theorists believe that substance use and abuse arises from a history of observing and partaking in addictive behaviors. The drug-using behavior then increases in frequency, duration, and intensity for the perceived psychological benefits it produces (Miller and McCrady, 1976). For some individuals, drinking can occur as part of growing up in a particular culture where the social influences of family and peers shape the behavior, beliefs, and expectations of the younger individuals concerning addiction (Monti, Abrams, Kadden, and Cooney, 1989). These theorists further explain that drinking and drug use is an acquired behavior that is maintained by reinforcement, modeling, conditioning, responding, expectations about alcohol effects, and physical dependence.

Behavioral theory

Thomb's (1994) research provides the counselor with a brief explanation of behavioral theory. Addiction is a behavioral disorder; it is learned. Like learning to drive a car, substance abuse is considered a behavior subject to the same principles of learning. It is seen as a problem behavior that is clearly under the control of environment, family, social, and/or cognitive contingencies. The addict is seen as a victim of a destructive learning pattern. According to the behaviorist, the use of alcohol or drugs is related to availability, lack of reinforcement for alternative behaviors, and lack of punishment for experimenting with substances.

Biopsychosocial model

The biopsychosocial model is a fairly new model of addiction that began to emerge in the late 1980s (Pratsinak and Alexander, 1992). This model sees addiction as a complex, progressive behavior having biological, psychological, sociological, and behavioral components (Donovan, 1988). According to this approach, multiple systems interact both in the development of addictive behaviors and in the treatment (Lewis, Dana, and Blevins, 1994).

The biopsychosocial model provides a metatheoretical framework that assumes the individual's total addictive experience comes from a web of interrelated and interdependent agents including biological, pharmacological, psychological, situational, and social components (Pratsinak and Alexander, 1992).

Alcoholics Anonymous (AA)

Alcoholics Anonymous developed to help alcoholics stop drinking and stay sober (AA World Services, Inc., 1976). Historically, there have been as many so-called treatments for addiction as there are theories to explain addiction, but one of the most recognized programs in the world has been AA. Alcoholics Anonymous is closely aligned with the medical/disease models of addiction (Pratsinak and Alexander, 1992).

Historically, AA is said to have been founded on June 10, 1935, the day that an alcoholic physician, Dr. Robert Holbrook Smith, had his last drink (Nace, 1987):

> At the hospital I humbly offered myself to God, as I then understood Him, to do with me as He would. I placed myself unreservedly under His care and direction. I admitted for the first time that of myself I was nothing; that without Him I was lost. I ruthlessly faced my sins and because I was willing to have my new found Friend take them away, root and branch, I have not had a drink since. (AA World Services, 1976)

However, the foundation of AA is said to have been set earlier in 1935, during a meeting in Akron, Ohio, between Dr. Robert (Bob) Smith, a well-known surgeon, and William Griffith Wilson, a New York stockbroker (AA World Service, Inc., 1976). Bill W. and Dr. Bob S. never claimed to be experts on alcoholism; they were men who shared a common problem and came together to support each other (Al-Anon Family Groups, 1995). From this common desire to help each other they developed AA, which has been sending the message of support and sobriety to millions ever since.

Alcoholics Anonymous encourages its members to take sobriety *one day at a time;* it is also a program that freely promotes and encourages a spiritual component, an appeal to a Higher Power. It is a non-profit, voluntary support group of recovering alcoholics who meet regularly to help each other get sober and stay sober. An advantage of the AA program is the use and role of the sponsor. Each member of AA obtains an individual sponsor who acts as a support, leader, friend, counselor— whatever it takes to assist their AA comrade to stay sober.

It is estimated that there are over 87,000 AA groups in existence, in over 150 countries, with a membership estimated at more than 1.7 million people (Miller and McCrady, 1993). As many as 95 percent of in-patient addiction treatment programs in the United States incorporated ideas from AA at some level (Bristow-Braitman, 1995). AA is a self-help group that does not provide treatment; instead it offers support for recovering alcoholics in their rehabilitation process.

The Twelve Steps and the Twelve Traditions constitute the core philosophy and structure of the program. The Twelve Steps of AA are introduced with this sentence: "Here are the steps we took, which are suggested as a program of recovery" (AA World Services, 1981):

Step 1   We admitted we were powerless over alcohol—that our lives had become unmanageable.

Step 2   Came to believe that a power greater than ourselves could restore us to sanity.

Step 3   Made a decision to turn our will and our lives over to the care of God as we understood Him.

Step 4   Made a searching and fearless moral inventory of ourselves.

Step 5   Admitted to God, to ourselves, and to another human being the exact nature of our wrongs.

Step 6   Were entirely ready to have God remove all these defects of character.

Step 7   Humbly asked Him to remove our shortcomings.

Step 8   Made a list of all persons we had harmed, and became willing to make amends to them all.

Step 9   Made direct amends to such people wherever possible, except when to do so would injure them or others.

Step 10   Continued to take personal inventory, and when we were wrong promptly admitted it.

Step 11 Sought through prayer and meditation to improve our conscious contact with God as we understand Him, praying only for knowledge of His will for us and the power to carry that out.

Step 12 Having had a spiritual awakening as the result of these steps, we tried to carry this message to alcoholics, and to practice these principles in all our affairs.

As the twelve steps became more broadly known, AA grew. This growth necessitated guidelines for the interrelationships among groups, and hence the Twelve Traditions of AA groups were developed (AA World Services, 1981):

Tradition 1 Our common welfare should come first; personal recovery depends upon AA unity.

Tradition 2 For our group purpose there is but one ultimate authority—a loving God as He may express Himself in our group conscience. Our leaders are but trusted servants; they do not govern.

Tradition 3 The only requirement for AA members is a desire to stop drinking.

Tradition 4 Each group should be autonomous except in manners affecting other groups or AA as a whole.

Tradition 5 Each group has but one primary purpose—to carry its message to the alcoholics that still suffer.

Tradition 6 An AA group ought never endorse, finance, or lend the AA name to any related facility or outside enterprise, lest problems of money, property, and prestige divert us from our primary purpose.

Tradition 7 Every AA group ought to be fully self-supporting, declining outside contribution.

Tradition 8 AA should remain forever nonprofessional, but our service centers may employ special workers.

Tradition 9 AA, as such, ought never be organized; but we may create service boards or committees directly responsible to those they serve.

Tradition 10 AA has no opinion on outside issues; hence the AA name ought not be drawn into public controversy.

Tradition 11 Our public relations policy is based on attraction rather than promotion; we need always maintain personal anonymity at the level of press, radio, and films.

Tradition 12 Anonymity is the spiritual foundation of all our traditions, ever reminding us to place principles before personalities.

Narcotics Anonymous (NA) was established in 1953. It is a self-help group patterned after AA (Lewis, Dana, and Blevins, 1994). The members of NA focus, not on any specific chemical, but on the disease of addiction. Narcotics Anonymous will accept members who are addicted to alcohol as well as narcotics.

Because substance abuse is a family issue, a need became apparent for support of co-dependent family members. Lois Wilson, wife of one of the co-founders of AA, became involved in an organization that in 1951 developed into *Al-Anon*. Al-Anon modified AA's twelve steps and traditions to meet the needs of non-alcoholic family members (Al-Anon Family Group, 1995). In 1957, Al-Anon established a group for teens, *Alateen,* with the goal of providing an opportunity for teenagers to come together to share experiences, discuss problems, learn how to cope more effectively, and provide encouragement to each other (Al-Anon Family Groups, 1969).

A PARADIGM TO ADDRESS ADDICTIONS

The Addiction Intervention Association (AIA) is the best-recognized association in Canada for the development, training, and preparation of addictions counselors. It is associated with the International Certification Reciprocity Consortium/Alcohol and Other Drug Abuse, Inc. (ICRC/AODA), which has put together twelve core functions that act as an outline for a rigorous certification program for people interested in addictions counseling. I have included the twelve core functions, the paradigm that ICRC/AODA teaches and promotes, as a core approach to work with potential addiction concerns. I recommend them as a practical and efficient philosophy for dealing with anyone suffering from addiction.

The *twelve core functions* for addictions counselors are as follows:

1. *Screening.* It must be determined if the client is eligible for admission based on the focus of the program, the target population, and funding requirements. It also must be determined if the applicant's drug or alcohol use constitutes abuse and if the program is appropriate to treat the abuse.

2. *Intake.* This is the completion of various forms, documents, and financial arrangements required for admission.

3. *Orientation.* This is the point where the counselor describes to the client the goals, rules, the hours services are provided, treatment costs, and the client's rights.

4. *Assessment.* Identification of the person's strengths, weaknesses, relevant history, and collaborative background information from people associated with the addicted individual. The results from the assessment provide focus for the development of a treatment plan.

5. *Treatment planning.* The counselor goes over the assessment results with the client and ranks the problems that are the most important to work on in his or her own view. Formulation of short-term and long-term goals.

6. *Counseling.* Counseling should focus on utilizing the strengths of the client in achieving the goals set out in the treatment plan. It is important to mobilize the client's existing resources.

7. *Case management.* Coordination of services is designed to put treatment goals into action.

8. *Crisis intervention.* The counselor should assess any crisis or problem that must be dealt with immediately.

9. *Client education.* Educate clients about alcohol and drug use/misuse as well as provide information about abuse issues (e.g., effects on the family, services and resources available).

10. *Referral.* Be aware of other resources that may be helpful to your client and be able to identify whether you as a counselor are not meeting your clients needs. In these instances, assistance from other sources may be necessary.

11. *Report and record keeping.* Document progress made in treatment and progress in attaining goals.

12. *Consultation.* When necessary, meet and discuss with other professionals client cases and progress.

Intervention and Techniques for Substance Abuse

When designing intervention programs it is important to establish treatment goals. The client must understand that internal change is vital to the recovery process. Clients must also be educated on the effects drugs have on their bodies and their life in general. Finally, substance abusers must remember that their addictions cannot be cured, only controlled, through their actions. Beating an addiction is not easy and relapse is common; however, help will always be available (O'Connell, 1990).

Drug Classification Table

The purpose of this section is to provide a basic introduction to the signs and symptoms of drugs commonly available and popular. The drug classification table, Table 25–1, is by no means a complete reference; its purpose is merely to act as a guide. We will undertake a very brief discussion of each drug category presented.

Cannabinoids

Restak (1994) explains that cannabinoids come from the *Cannabis sativa* plant, which has more than 400 compounds, of which 61 have psychoactive properties. The main psychoactive ingredient in cannabinoids is THC. Cannabinoids do not fit under any other category because they generally act as a depressant but sometimes act as a stimulant (e.g., elevation in heart rate).

Hallucinogens

Hallucinogens are drugs that affect an individual's perception, affect, and cognitive processes. Users may believe they have a new insight into reality (Snyder, 1986).

Mixed action drugs

Mixed action drugs are so named because their effects depend on the dose and route of administration; thus, they can act as an anaesthetic, stimulant, depressant, or hallucinogen (Brown and Braden, 1987). Phencyclidine (PCP, or angel dust) is classified as a mixed action chemical.

Table 25-1.   Drug Classification (adapted from Health and Welfare Canada, 1990)

CANNABINOIDS

Name	Description	Short-Term Effects	Long-Term Effects	Tolerance and Dependence
Marijuana, pot, weed, joint	Flower top or leaves range in color from green to brown with a coarse texture. Smoked in pipe or cigarette	Effects of smoke felt in a few minutes. High lasts from 2 to 4 hours. User feels calm, relaxed, appetite increases, can cause psuedo hallucinations. Some users experience anxiety and depression. Known to effect motor skills. Very dangerous to drive when taking	Can result in serious brain damage known as *amotive syndrome*. Decreases motivation, concentration. Lack of interest in the world, unrealistic thinking, impaired communication, general apathy	Psychological dependence. DSM-IV (1994) explains that physical dependence does not usually develop. Research now exploring this question
Hashish, hash	Dried resin from cannabis plant, usually brown to black. Smoked in pipe, cigarette, alone or in baked goods			
Hash oil, oil	Thick greenish-black or reddish brown obtained from mixing hash with an organic solvent.			

HALLUCINOGENS

Name	Description	Short-Term Effects	Long-Term Effects	Tolerance and Dependence
LSD (Lysergic acid diethylamide)	Sold on blotting paper, gelatin paper, as tablets, capsules or liquid solution. Common dose 50–100 mcg taken orally	Effects within an hour, high lasts from 2 to 12 hours. Possible effects on perception, colors brighter, extreme mood swings, joy to depression, fear, and terror	Decreases motivation and increases depression and anxiety. LSD highs may spontaneously occur months or weeks later (flashbacks)	Psychological tolerance occurs but does not cause physical dependence
Mescaline or peyote (trimethoxy-phrene-thalamine)	Found in powders, tablets, or capsules. Usually taken orally in doses of 300–500 mg	Effects appear slowly, last 10–18 hours. Changes in perception and mood, disorientation	Insufficient research	Psychological tolerance, crosses over with LSD. No physical dependence

HALLUCINOGENS (continued)

Name	Description	Short-Term Effects	Long-Term Effects	Tolerance and Dependence
Psilocybin (magic mushrooms)	Mushrooms, or in capsules taken orally in doses of 5–60 mg	Effects around 1 hour, last for several hours. Sensation of relaxation, separation of surroundings. Large doses cause dizziness, discomfort, nausea, flushing, and sweating	Insufficient research	Psychological tolerance, crosses over with mescaline. Not known to cause physical dependence
Other hallucinogens that the professional may want to research	1. MDA (methylenedioxyamphetamine) 2. STP or DOM (2.5-dimethoxy-4-methylamphetamine) +			

MIXED ACTION

Name	Description	Short-Term Effects	Long-Term Effects	Tolerance and Dependence
PCP (phencyclidine, angel dust). Used as an alternative for many illegal drugs, e.g., LSD.	Sold as powder of any color, in crystal, liquid, tablet, capsule or paste. Can be smoked, sniffed, swallowed, or injected (1–5 mg enough for a high)	Effects last 3–18 hours. Can produce a state of pleasure, sense of separation. However, many bad trips occur that lead to aggressive, paranoid, and confused behaviors	Flashbacks, speech problems, severe psychological problems	Psychological tolerance occurs, but no physical dependence

NARCOTICS, OPIATES

Name	Description	Short-Term Effects	Long-Term Effects	Tolerance and Dependence
Opium	Dark brown chunks or powered, eaten or smoked	When injected, users feel state of gratification. Body feels warm, dry mouth. User goes into a stupor (when taken orally, effects slower). Physical effects can cause vomiting, contraction of pupils, skin cold, breathing slows, can go to complete stop (death)	Constipation, moodiness, lung problems, AIDS (dirty needles and injection equipment), organic brain syndrome, complications during childbirth, liver damage. During withdrawal, users may take painkiller to help with withdrawal, and may become tolerant	Tolerance builds quickly, resulting in severe phyiscal dependence. Withdrawal lasts 7–10 days. Severe withdrawal symptoms: severe anxiety, sweating, muscle spasms, chills occur within 5 hours of last dose

NARCOTICS, OPIATES (*continued*)

Name	Description	Short-Term Effects	Long-Term Effects	Tolerance and Dependence
Codeine	Tablets, capsules, solutions			
Morphine	Injections and tablets			
Heroin diacetyl morphine, H, horse, smack	Fine white or brown powder, can be sniffed usually injected			
Methadone® (® denotes prescription drugs)	Soluble powder injected or aken orally			
Others to be aware of:	Demerol® Dilaudid® Novahistex-DX® Percodan® Talwin ® Lomotil® MS Contin ®			

INHALANTS/SOLVENTS

Name	Description	Short-Term Effects	Long-Term Effects	Tolerance and Dependence
Volatile organics	Fuel in glues, gas, paint thinners, and cleaning agents	Lightheadedness, euphoria, fantasy, recklessness. Inhalants enter body via bloodstream from lungs. Decreases respirations and heart-rate. Potential risk of sudden death	Feelings of weightlessness, fatigue, hostility, paranoia. Kidney and liver damage. Increased blood-lead levels can cause brain damage	Regular use leads to tolerance. Both physical and psychological dependence develops. Withdrawal symptoms include chills, headache, cramps, tremors

CENTRAL NERVOUS SYSTEM DEPRESSANTS

Name	Description	Short-Term Effects	Long-Term Effects	Tolerance and Dependence
Alcohol (ethyl alcohol or ethanol)	Pure alcohol is a clear substance	Affects brain and spinal cord. Small doses affect reason, caution, memory. Increased doses affect senses, judgement, coordination, and balance. Larger doses affect vital centers, e.g., respiration, which can lead to death	Damages pancreas, liver, heart, stomach, intestines, increases risk of certain cancers, e.g., esophagus, leads to eventual brain damage—Wernicke syndrome and Korsakoff syndrome	Regular use leads to both physical and psychological dependence. Tolerance increases with use. Withdrawal ranges from sweating, hallucinations, vomiting, delirium tremens to death
Barbiturates Secobarbital ®, reds; phenobarbital ®, yellows; amobarbital ®, blue heaven. Barbiturate replacements: methaqualone ®, flurazepam® , triazolam ®, Halcion	Drugs prescribed by a physician for medical purposes	Affect central nervous system, e.g., depress respirations, slurred speech, perceptions. Overdose may lead to death	Liver damage, chronic intoxication, depression, birth defects (e.g., baby has difficulty breathing and sleeping)	Regular use leads to tolerance, crosses over with alcohol. Psychological and physical dependence occurs. Withdrawal similar to alcohol
Benzodiazepines (minor tranquilizers, nerve pills) Valium®, diazepam; Librium®; chlordiazepoxide; Serax®, oxazepam; Ativan®, lorazepam; Xanax ®, alprazolam	All of these tranquilizers are prescribed by physicians for medical purposes and have markings on the tablets. To identify a marked tablet, take one to a pharmacy; they can tell the drug's origin and classification and provide detailed information on the drug	Relaxed muscles, reduced mental alertness (should not drive a motor vehicle), feeling of well-being. Side effects of skin rash, nausea, dizziness. Very dangerous if mixed with alcohol or other sedatives	May lead to increased aggressiveness. Some of these drugs accumulate in the body, thereby increasing their effect. Babies born from user mothers may experience withdrawal symptoms	Tolerance increases as psychological and physical dependence occurs. Withdrawal: tremors, cramps, sweating, vomiting and even death. Because signs of addiction are not clearly evident, some believe the drug is harmless

CENTRAL NERVOUS SYSTEM STIMULANTS

Name	Description	Short-Term Effects	Long-Term Effects	Tolerance and Dependence
Cocaine, crack, coke, snow	Fine, white crystalline powder that can be put into different forms. Smoked high: 7–8 sec. Injected: 15 sec. Snorted: 3 min. Eaten: 10 min.	Feeling of euphoria, energetic, alert, rapid heart beat and respirations. Large doses cause violence, tremors, hallucinations, chest pain and death. Danger of impurities used to cut cocaine may cause fatal allergies	Malnutrition, psychosis, prone to violence, risk of AIDS, kidney damage, liver and tissue damage. Babies born to user mothers risk withdrawal symptoms	Cocaine produces a very powerful psychological addiction. Physical dependence occurs as well. Withdrawal symptoms may include sleep disturbance, fatigue, hunger, depression, and violence
Amphetamines: Dexadrine®, dextroamphetamine; Methadrine® (methamphetamine, or "ice"); Tenuate®, diethylpropion; Ionamin®, phenteramine; Ritalin®, methylphenidate; Fastin®, phenteramine	Produced for and prescribed by physicians, for treatment of medical conditions	Increased alertness and energy, feeling of well-being. Elevated blood pressure, heart and respiratory rate, sweating, dilated pupils, talkative, restless, aggressive. Large doses: pallor, rapid, irregular heart ate, tremors, severe paranoia, hallucinations. Death can result from ruptured blood vessels in brain. Violence is the leading cause of amphetamine-related deaths	Malnutrition, amphetamine psychosis. Prone to violence. AIDS risk and other infections from repeated injections, damage to blood vessels. Kidney damage, lung problems, stroke, and other tissue damage. Newborns will experience withdrawal symptoms	Powerful psychological dependence, physical dependence may also occur. Withdrawal symptoms include sleep disturbance, fatigue, long but disturbed sleep, strong hunger, irritability, depression, violence
Nicotine	Shredded leaves of tobacco plant. Contains tar, nicotine, carbon monoxide, and 400 other dangerous compounds. Is smoked or chewed	Increased heart and respiratory rate, blood pressure. Decreased appetite and skin temperature	Each cigarette smoked cuts 5.5 min from lifespan. Predisposes to various cancers: mouth, throat, lung, heart, and/or respiratory disease	Physical and psychological dependence. On quitting, some damage may not be reversible

CENTRAL NERVOUS SYSTEM STIMULANTS (*continued*)

Name	*Description*	*Short-Term Effects*	*Long-Term Effects*	*Tolerance and Dependence*
Caffeine	Bitter white substance derived from coffee bean, tea leaves, cocoa leaves, kola nut. Available as coffee, tea, chocolate, cola drinks, and in many medications	Increased metabolic rate, blood pressure, body temperature, stomach acid. Decreases sleep, appetite, fine coordination (tremor). Large doses (10 gm orally) produces headache, nervousness, and delirium	Regular doses of 8 cups/day produce chronic insomnia, persistent anxiety, depression, and stomach upset. Linked with birth problems	Physical and psychologic dependence occurs. Withdrawal symptoms include severe headache, irritability, and fatigue

PERFORMANCE ENHANCERS

Name	*Description*	*Short-Term Effects*	*Long-Term Effects*	*Tolerance and Dependence*
Anabolic steriods Abused by both athletes and non-athletes (to build self-confidence)	Found in oral tablets or injected	Behavior changes, mood swings, change in sleep patterns, weight gain and strength gains, anxiety and depression, risk of AIDs with needle use	Liver damage, tendon damage, heart attacks, strokes, diabetes, hair loss, infertility in males	Psychological dependence

Narcotics and opiates

Narcotics and opiates are analgesic drugs that relieve pain. They are produced from the opium poppy plant (*Papauer somiferum*), or synthetically.

Inhalants and solvents

Inhalants and solvents are a diverse group of toxic substances that have been manufactured for industrial use. When these substances are inhaled, they will alter brain function.

Central nervous system depressants

Central nervous system depressants act on the body as a sedative or hypnotic. All of the drugs in this group have considerable abuse potential, and are both physically and psychologically addictive.

Central nervous system stimulants

Central nervous system stimulants are drugs that stimulate the central nervous system, producing a quick, short-term increase in energy. This group of drugs can lead to severe physical and psychological dependance.

Performance enhancing drugs

Anabolic steroids, popular for enhancing performance, act on body tissues to increase the speed of growth by stimulating protein synthesis, which may increase muscle size and strength (Gottesman, 1992). Their use has unpredictable outcomes, including death. Doweiko (1996) explains that these drugs, when abused, may also induce a sense of euphoria; however, this is not the main reason for use.

Assessment

This section covers only a few of the instruments and tools available for the assessment of substance abuse. If you are interested in learning more, we recommend a review of the literature, or contact your local agency on drug dependency for information on tools currently favored.

Diagnostic and Statistical Manual of Mental Disorders, 4th edition (DSM-IV)

The DSM-IV (1994) is an instrument developed, and subsequently revised, by the American Psychiatric Association (APA). We have reviewed its relevant sections earlier in this chapter.

The Michigan Alcohol Screening Test (MAST)

Lewis, Dana, and Blevins (1994) identify the MAST as an assessment instrument that is made up of 25 questions regarding the client's involvement with drinking. The respondent answers yes or no. MAST scores range from 0 to 53. The respondent's total score reflects the degree to which drinking is a problem. The MAST scale is:

0   no problem; no action

1 or 2   low degree of problems; monitor client for now and reassess at a later time

3 to 5   moderate level of problems; further investigation

6 to 8   substantial level of problems; intensive assessment required

9 or 10   severe level of problems; intensive assessment required

CAGE Drinking Questionnaire

The CAGE Drinking Questionnaire asks four questions about respondents' behaviors and thoughts about their drinking and how it has affected their life. If the respondent gives two or more positive responses, there is sufficient evidence of alcohol abuse; further investigation is warranted. Evans and Sullivan (1995) provides the following CAGE questions:

Cut down. Have you ever tried to cut down your use of chemicals?

Annoyed. Has anyone ever annoyed you by criticizing your drinking?

Guilty. Have you ever felt guilty about your behavior when drinking?

Eye opener. Have you ever used alcohol in the morning to reduce the effects of a hangover?

Addiction Severity Index

Lewis, Dana, and Blevins (1994) explain that this is one of the more reliable instruments that assess dependency of drugs other than alcohol. It explores several areas: medical status, employment status, family and social relationships, and psychological status. The Addiction Severity Index uses a detailed interview, with each question being assessed on a continuum ranging from 0 (no concern) to 9 (treatment needed/life threatening situation).

Substance Abuse Subtle Screening Inventory (SASSI)

The SASSI is a comprehensive instrument used to uncover chemical dependence; it is considered to be reliable. It uses both true/false and 0–4 continuum questions to determine the client's dependence. This instrument also considers gender profiles, in that it provides separate scoring sheets for males and females. It is an excellent instrument. For training in this instrument, contact the SASSI Institute at 1-800-726-0526.

Dual Diagnosis Treatment Strategies

Sometimes patients in a treatment program have a coexisting mental disorder in addition to their substance abuse problem (e.g., alcohol dependence plus schizophrenia). The counselor must be aware of the proper diagnosis of the illness in order to plan treatment. Note that the *human services counselors are not qualified to make such diagnoses* and will need to rely on professional assistance. However, you may serve as part of a multidisciplinary team and thus be involved in treatment planning. In some cases, an experienced human services counselor with advanced education may be able to make accurate assessment that will then be endorsed by the professional community.

One tool a human services counselor *can* use, the decision tree, is a way of identifying special treatment needs and defining treatment priorities for the dual diagnosis client. The decision tree presented next is an adaptation of Wolfe-Reeves (1990). It is used for assessing psychiatric symptoms before creating treatment plans for the chemically dependent.

Decision Tree
To assess for potential violence:

1. Evaluation of violent expressions in the acutely intoxicated individual:
   - Symptoms indicating poor reality testing or poor impulse control (e.g., to prevent suicides)
   - History of violence
   - History of mental illness
   - Secondary gain
   - Available support systems

2. Evaluation of violence after detoxification:
   - [Client] expresses regret or amnesia
   - [Counselor] uses confrontation to prevent recurrences and to stress patient's responsibility for actions while under the influence

3. Evaluation of violence in non-intoxicated individuals:
   - Symptoms indicating poor reality testing
   - Seriousness of intent
   - History of violence
   - History of mental illness
   - Situational factors (precipitating events)
   - Treatment context

To assess risk of medical danger:

1. Assess for signs and symptoms of delirium (e.g., hallucinations).

2. Do a history of recent events:
   - Assess symptoms that may interfere with the client's ability to participate in treatment for addictions. These may include: isolation, withdrawal or difficulty relating to others—extremes of mood, mood swings, a labile mood—halucinations, delusions, poor memory, limited attention span, or poor concentration.
   - Evaluate additional factors that may influence the assessment process and the outcome of treatment—things such as the attitude of the evaluator, situational factors, treatment options, and the availability of assessment tools.

Studies show that 1 in 3 who seek help for a chemical dependency problem are likely to satisfy the criteria for another psychiatric disorder. Dually diagnosed patients have a poorer prognosis than those who have a single substance use disorder. These patients respond best when their multiple problems are addressed in a treatment plan suited to their individual needs (Wolfe-Reeves, 1990).

Carroll (1990) states three considerations when treating a dually diagnosed client:

1. Clients with mental disorders can achieve sobriety.

2. The proper use of prescribed psychotropic medication is essential.

3. A flexible treatment strategy geared to the individual's specific needs is the key to success when working with this population. In addition, these clients must have an aftercare program that monitors both their substance abuse recovery and their mental disorder.

Covert (or Verbal) Aversion Therapy

Lewis, Dana, and Blevins (1994) describe aversion therapy as a treatment program that attempts to stop the substance abuse by creating an aversion or distaste for the drug. Example:

Step 1   Counselor teaches relaxation responses to the client.

Step 2   Client imagines aversive pictures involving the drug and then repeatedly recalls these pictures with images of the sight, smell, and taste of the drug.

Step 3   With repetition of these paired images, client will have imagined aversion consequences for taking the drug. This eventually leads to an aversion to the abused substance.

The Action Counseling Model

George (1990) explains action counseling as a three-stage model emphasizing skills that will assist clients to take control of their life.

Stage 1   The counselor works to establish a therapeutic relationship with the client, gathers pertinent information about the client's history and information about the client's present difficulties. In this stage the counselor becomes aware of, and learns about, the client's world.

Stage 2   The most important outcome of this stage is that the client determines the necessary changes, sets goals for change, and then makes the *commitment* to change.

Stage 3   In this stage, clients begin to take action to bring about the desired change in their life. The key component of the action stage is that the client make a commitment to the direction of desired change. The client, with the help of the counselor, develops strategies and begins to put them to work (e.g., a behavior contract is agreed to that the client will attend counseling twice a week for the next month).

Behavioral Contracts

With behavioral contracts, clients state in advance the type of behavior they want to change and then hold to that commitment through preparing the contract. Clients tend to take written contracts more seriously than oral ones. The contract should be simple and nonjudgmental and be signed by both client and counselor. Once the contract is finalized, it can be put on a small card called the *Stay Safe Card* that contains the main points of the contract, a plan to stay safe, and a phone number for help.

When clients experience a crisis moment where they are tempted to breach the contract, I suggest you go right to your sex-step crisis intervention model that was introduced in Chapter 20. The goal is to get the client safe and stabilized and back on plan. Find the behaviors that were the triggers for the lapse and ensure that the plan addresses these issues.

Minnesota Model of Chemical Dependency

Doweiko (1996) explains that the Minnesota model is one of the more successful models in addressing addiction problems with drugs or alcohol.

Stage 1   *Evaluation.* Members of the treatment team meet with the client to assess and make recommendations with regard to the client's needs.

Stage 2   *Goal setting.* The treatment team then meets and recommendations are made to the client's case manager (who is usually the client's chemical counselor). The client, family, and others the client interacts with, can participate at the treatment planning meeting, and are free to make recommendations. From this meeting the recommended and appropriate treatment plan is agreed upon by the team.

Stage 3   *Developing a formal treatment plan.* The case manager and the client enter this third stage together. The strength of the Minnesota model is that it allows a number of people to be involved directly with the client's recovery. The treatment plan includes the following: (a) the client's specific problem areas are identified (medical, psychological, social, recreational, or spiritual), (b) behavioral objectives are outlined (short-term and long-term), (c) ways to measure progress are established, and (d) a target date is set for each goal.

Harm Reduction Model

Marlatt (1994) describes the harm reduction model as being based on moderation instead of the traditional method of abstinence. It focuses on reducing the degree of damage caused by the addiction until clients are able to control their life or choose abstinence. Example:

• For clients addicted to cigarette smoking, the nicotine skin patch is used. The harm of the actual smoking process is reduced, and clients can slowly be weaned off the need for nicotine at a rate they are able to handle.

Relapse Prevention

CENAPS model

I have chosen the CENAPS (Center for Applied Sciences) model developed by Terence Gorski because it is based on the biopsychosocial model of addictive disease. Chemical addiction causes brain dysfunction that disorganizes personality and causes social and occupational problems. Total abstinence as well as personality and lifestyle changes are essential for full recovery. A number of principles are associated with the CENAPS model of relapse prevention therapy. Each principle forms the basis of specific relapse prevention therapy procedures.

Principle 1 *Self-regulation*

- Procedure: Stabilization through detoxification. There is a five-day "dry out" program and 28- day detox program to establish some daily structure for the client. It involves working on diet, exercise, stress management, medical intervention, counseling, meetings for NA/AA, and so on.

Principle 2 *Understanding the recovery relapse stages*

- Procedure A: Self-assessment is done by clients with their increasing awareness of the recovery process, including warning signs of relapse and high-risk situations.

- Procedure B: Relapse education sessions are set up to provide clients with information about recovery, relapse, and prevention planning.

Principle 3 *Identification of relapse warning signs*

- Procedure: A general description of the signs should be put on an emergency sobriety card to assist the individual in identifying it in a high-risk situation.

Principle 4 *Management of relapse warning signs*

- Procedure: Warning-signs management is done through mental rehearsal, role playing, and therapeutic assignments.

Principle 5 *Recovery plan*

- Procedure: Develop a recovery plan including warning signs, small steps, family time, exercise, diet, AA/NA meetings, counseling, work, hobbies, daily journals, management methods, and a sobriety card.

Principle 6 *Daily inventory*

- Procedure: Take things one day at a time and keep a log of how the plan is working. At the end of the day, review progress, discuss results, and troubleshoot the day.

Principle 7 *Significant other/family involvement*

- Procedure: Family involvement. Loved ones are brought into treatment when appropriate. Services and education should be provided.

Principle 8 *Relapse prevention plan update*

- Procedure: Within the first six months, relapse is most likely to occur. Follow up after initial care to update treatment plan.

Principle 9   *Thinking in principles*

- Procedure: Requires the identification of core issues and mistaken beliefs that create the irrational thinking, unmanageable feelings, and self-destructive behavior that result in dysfunction during recovery.

Relapse Educational Model (Daley, 1991)

It is important to educate the client on ways to detect early signs of relapse. This intervention can be used in combination with other intervention models or as a sole means of prevention.

Session 1	Understanding the relapse process
Sessions 2 and 3	Identifying and handling high-risk situations
Session 4	Identifying and handling urges or cravings
Session 5	Identifying and handling social pressure
Session 6	Anger management
Session 7	Handling boredom and using leisure time
Session 8	Stopping actual relapse

Other Recommended Interventions for Substance Abuse

The reader is referred to other chapters for the following:

- Roger's core conditions (see Chapter 8)
- Confronting (see Chapter 16)
- Relaxation training (see Chapter 21)
- Modeling (see Chapter 18)
- Social skills training (see Chapter 11)
- Assertion training/refusal skills (see Chapter 18)
- Cognitive restructuring (see Chapter 16)
- Systematic desensitization (see Chapter 18)
- Spirituality (see Chapter 9)
- Activity scheduling (see Chapter 18)

INTERVENTIONS FOR PATHOLOGICAL GAMBLING

Heineman (1988) explains that people who start down the road to a gambling addiction go through three stages: winning, losing, and desperation. We have hypothesized that, after these three stages, a gambler must go through four more stages (self-evaluation, education and support, practicing and implementation, and new beginning) to reach a point of recovery or "path to sanity" (Figure 25–1). But we have chosen to start with what we term the *pre-gambling* phase.

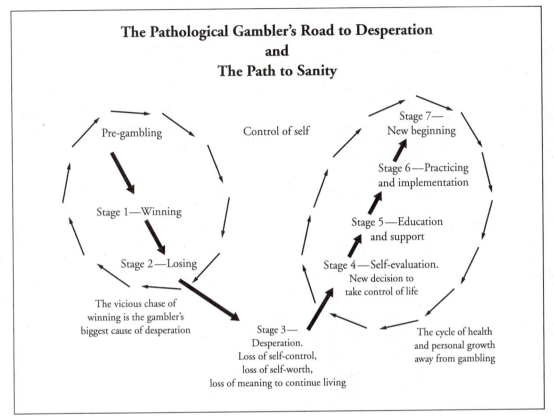

Figure 25–1.

Eclectic model for change (adapted from Heineman's *Three Stages to Pathological Gambling,* 1988)

Pre-Gambling

The pre-gambling phase is defined as the time when a person is not yet a compulsive gambler but exhibits the potential characteristics (low self-esteem, hopelessness, boredom) to become one. Once the first bet is made, this individual has started down the road to desperation.

Many gamblers were gamblers before they laid their first bet. When they are introduced to gambling, they perceive it as an option to meet their unmet needs. Given certain personal attributes (e.g., a lack of self-control), they may eventually be involved in a vicious chase to win. Thus we begin to see the three stages of Heineman (1988).

Stage 1 *Winning phase.* Typically, in this phase gambling is sporadic, with some positive rewards. In other words, it is fun. "The money set aside for gambling is similar to money spent at an amusement park: a specific amount of discretionary money is designated for this leisure activity, which becomes a happy memory" (Heineman, 1988). This leisure activity is accepted by family members because they share the rewards (psychological or financial) of the gambler's so-called good luck. "In this phase, most gamblers experience their first big win—usually equivalent to several months' income" (Heineman, 1988). The gambler's losses in this phase are tolerable when compared to winnings.

Gamblers in this winning phase reassure the family, deny there is a problem, and let those around them believe that everything is under control and there is no cause for concern. The gambler's repeated response is "Don't worry about it" (Heineman, 1988).

Stage 2 *Losing phase.* "The losing phase begins whenever the gambler increases the amount and frequency of the gambling and gambles compulsively" (Heineman, 1988). The losses outweigh the winnings for the first time. In this phase, family members begin to worry as family finances are negatively affected. The lies that are characteristic of this illness add to the anxiety of individuals in the gambler's family. Before long, any trust the family had in the gambler is broken. During the losing phase, the compulsive gambler:

- Experiences prolonged losing episodes
- Thinks continually about gambling
- Becomes careless about the welfare of loved ones
- Tries to cover up the gambling
- Gets into heavy legal and illegal borrowing
- Turns to friends and family members for bailouts in order to have the money to continue gambling (Heineman, 1988)

In this phase, there is a noticeable difference in the gambler's personality. The gambler feels both anxious and excited when gambling. The anxiety comes from the fear of losing; the excitement comes from the gambler's abnormal belief: "I am special. Lady Luck is on my side. I can't lose" (Heineman, 1988). In this phase, gamblers continue to search for bailouts to cover losses that will enable them to place another bet. The cycle then begins again.

Stage 3 *Desperation phase.* This is the phase where the gambler's main focus is on ways to come up with the money to support the gambling habit. "By now, the gambler has probably resorted to illegal ways of obtaining money, such as through loan sharking, stealing, or embezzlement. Fear, frustration, anger, and depression become the first order of each and every day" (Heineman, 1988). In this phase, gambling has taken over the person and it is no longer a choice. The act of gambling is now the gambler's fundamental concern.

Stage 4 *Self-evaluation.* In this stage, gamblers recognize their destructive behavior and make the decision to take control of their life. They willingly admit to themselves and others (family member, friend) that they have the compulsion to gamble. This sometimes happens when gamblers hit "rock bottom" and realize their behavior is destroying their own lives and those of others. They want to stop gambling.

Stage 5 *Education and support.* The gambler, at this stage, has reached out for support through family, friends, and/or therapy. The gambler is educated about the compulsion to gamble and taught ways to recognize the warning signs that lead to this compulsive behavior. Through the educational process, the gambler learns ways to modify the behavior. Family and friends are introduced to *Gam-Anon* to help to understand compulsive gambling and provide support to the gambler when needed. Also in this stage, gamblers will be working on the characteristics that carried them from the pre-gambling stage to the compulsion to gamble (e.g., low self-esteem, hopelessness). This can be done through various methods such as therapy and self-help groups.

Stage 6 *Practicing and implementation.* In this stage, the gambler is practicing and implementing all that has been learned in stage 5. The gambler will have a support system (e.g., family, friends, GA) for encouragement and support through recovery. The gambler is able to recognize the signs and triggers that set off the compulsion to gamble, but now has the coping skills and support systems to help overcome it.

Stage 7 *New beginning.* Once people have reached this stage they have established a newfound "quality of life," a state of being without gambling controlling them. The person has worked (and will continually work) on the personal issues that were present in the pre-gambling stage. *Note:* For ex-gamblers to continue living a "quality life," they will continually be revisiting the last three stages (the cycle of health) to ensure that they do not fall back into gambling as well as to assist them in reaching their full potential as happy, healthy human beings.

Diagnostic Criteria for Gambling

It is important for a counselor who works with addicted individuals to understand pathological gambling, as well as dual diagnosis issues (e.g., substance abuse plus gambling). The DSM-IV (1994) defines *pathological gambling* as follows:

A. Persistent and recurrent gambling behavior as indicated by five (or more) of the following:

   (1)   is preoccupied with gambling (e.g., preoccupied with reliving past gambling experiences, or planning the next venture, or thinking of ways to get money with which to gamble)

   (2)   needs to gamble with increasing amounts of money in order to achieve the desired excitement

   (3)   has repeated unsuccessful efforts to control, cut back, or stop gambling

   (4)   is restless or irritable when attempting to cut down or stop gambling

   (5)   gambles as a way of escaping from problems of relieving a dysphoric mood (e.g., feelings of helplessness, guilt, anxiety, depression)

   (6)   after losing money gambling, often returns another day to get even ("chasing one's losses")

(7)   lies to family members, therapist, or others to conceal the extent of involvement with gambling

(8)   has committed illegal acts such as forgery, fraud, theft, or embezzlement to finance gambling

(9)   has jeopardized or lost a significant relationship, job, or educational or career opportunity because of gambling

(10) relies on others to provide money to relieve a desperate financial situation caused by gambling

B. The gambling behavior is not better accounted for by a Manic Episode. (DSM-IV, 1994)

Gamblers Anonymous

Gamblers Anonymous (GA) is a group therapy technique that uses only ex-gamblers as helpers and is modeled on other self-help groups such as Alcoholics Anonymous (AA) (Scodel, 1964). Brown (1993) points out differences between GA and AA:

- GA meeting numbers are smaller than AA.
- GA meetings are usually longer than AA meetings.
- GA has about 20 weekly meetings, while AA has many more meetings per week, and throughout the day.
- *Gam-Anon* meetings are usually held on the same night and location as GA; Alanon meetings are not held at the same time and location as AA meetings.
- In GA, one "gives therapy," while in AA, one "shares."
- In GA, one "jumps," while in AA, one has a "slip" or returns to drinking.
- GA has a pressure relief group, provided by GA members, called Trusted Servants; [crisis] is referred to as "the moment of truth." AA does not have this.
- GA has relatively few step meetings compared to AA.
- GA conceptualizes the problem of addiction differently from AA.
- GA de-emphasizes God and spirituality compared to AA.
- GA is organized differently from AA.
- GA differs from AA in the nature of members' consciousness.
- GA rejects the conception of the "inner child" that is used in AA (Browne, 1993).

OTHER INTERVENTIONS

In conclusion, we will touch briefly on a number of other intervention strategies.

- *Structured family intervention.* This model initiates intervention with the individuals directly affected by the gambling (friends and family) and its objective is to assist the "gambler" to stop gambling. The gambler is provided with a support system having the necessary knowledge and skills (e.g., to ensure all the family's money is protected, because active gamblers will lie about how they use money). Friends and family can acquire the skills and understanding of the gambler's illness by attending Gam-Anon. Here they will receive support, compassion, and affirmation. The gambler is supported as a person; the habit is not supported (Heineman, 1988).

- *Cognitive behavior.* Cognitive therapies attempt to alter the thinking patterns of a gambler so that the unwanted behavior is no longer triggered.

- *Group psychotherapy.* Effective therapy for gamblers should involve changes in the maladaptive personality traits that lead them to the compulsion of gambling. To help the gambler deal with everyday problems (finances, relationships) the treatment must be tailored to the gambler's needs. Group psychotherapy usually includes both individual and marital therapy involving education on health and addiction.

- *Reality therapy* (Chapter 15). This is an excellent tool for helping people with addictions.

- *Behavior modification* (see Chapter 18).

Summary

This chapter presented an introduction to the complexity of the field of addictions. Most human services counselors will come in contact with a client who is self-medicating to the point of having an addiction of some sort. Thus, all human services counselors need to have an understanding of what an addiction is, and how it can be treated.

References

Al-Anon Family Groups. (1969). *Facts about Alateen.* New York: Author.

Al-Anon Family Groups. (1995). *How Al-Anon works for families and friends of alcoholics.* Toronto: Author.

Alcoholic Anonymous World Services, Inc. (1976). *Alcoholics anonymous* (3rd ed.). New York: Author.

Alcoholics Anonymous World Services, Inc. (1981). *Twelve steps and twelve traditions.* New York: Author.

American Psychiatric Association. (1994). *Diagnostic and statistical manual of mental disorders, Fourth Edition.* Washington, DC: Author.

Bristow-Braitman, A. (1995). Addiction recovery: Twelve steps program and cognitive-behavioral psychology. *Journal of Counseling and Development 73:*414–18.

Brown, R., Braden, N. (1987). Hallucinogens. *Paediatric Clinics of North America 34*(2):341–47.

Cloniger, C. R., Gohman, M., Sigvardsson, S. (1981). *Cross Archives of General Psychiatry 38:*861–68.

Daley, D. C. (1988). *Relapse, conceptual research and clinical perspectives.* Binghamton, NY: Haworth.

Donovan, D. M. (1994). Assessment of addictive behavior: Implication of an emerging biopsychosocial model. In Lewis, J. A., Dana, R. Q., Blevins, G. A., *Substance abuse counseling* (2nd ed.). Pacific Grove, CA: Brooks/Cole.

Doweiko, H. E. (1996). *Concepts of chemical dependency* (3rd ed). Pacific Grove, CA: Brooks/Cole.

Eadington, W. R., Cornelius, J. A. (1993). *Gambling behavior and problem gambling.* Reno, NV: University of Nevada.

Edwards, G., Arif A., Jaffe, J. (Eds.). (1983). *Drug use and misuse.* New York: St. Martin's Press.

Evans, K., Sullivan, J. M. (1995). *Treating addicted survivors of trauma.* New York: Guilford.

George, R. L. (1990). *Counseling the chemically dependent.* Needham Heights, MA: Allyn & Bacon.

Gold, M.S. (1991). *The good news about drugs and alcohol.* New York: Villard.

Heineman, M. (1988). *When someone you love gambles.* Center City, MN: Hazelden.

Jacobs, M. R., Fehr, K. (1987). *Drugs and drug abuse* (2nd ed). Toronto: Addiction Research Foundation.

Jellinek, E. M. (1960). *The disease concept of alcoholism.* New Haven: Yale University Press.

Jellinek, E. M. (1995). Alcohol, science, and society. Twenty-nine lectures with discussions as given at the Yale Summer School of Alcohol Studies. *Quarterly Journal of Studies on Alcohol 13:* 673–764.

Kissin, B., Begleiter, H.(Eds.). (1977). *The biology of alcoholism vs. the chronic alcoholic.* New York: Plenum.

Lester, D. (1980). The treatment of compulsive gambling. *International Journal of Addictions 15*(2):204.

Lewis, J. A., Dana, R. Q., Blevins, G. A. (1994). *Substance abuse counseling* (2nd ed.). Pacific Grove, CA: Brooks/Cole.

Monti, P. M., Abrams, D. B., Kadden, R. M., Cooney, N.L. (1989). *Treating alcohol dependency.* New York: Guilford.

O'Connell, D. F. (Ed.). (1990). *Managing the dually diagnosed patient.* New York: Haworth.

Payne, W. A., Hahn, D. B., Pinger, R. R. (1991). *Drugs: Issues for today.* St. Louis: Mosby.

Prashant, S. (1993). *Drug abuse and society.* New Delhi: Anish Publishing House.

Pratsinak, G., Alexander, R., (Eds.). (1992). *Understanding substance abuse.* Laurel, MD: Goodway Graphic.

Sappington, A. A. (1989). *Adjustment theory, research and personal application.* Pacific Grove, CA: Brooks/Cole.

Schuckit, M. A. (1989). *Drug and alcohol abuse: A clinical guide to diagnosis and treatment* (3rd ed). New York: Plenum.

Snyder, S. (1986). *Drugs and the brain.* New York: Scientific American.

Thombs, D. L. (1994). *Introduction to addictive behavior.* New York: Guilford.

Wilford, B. B. (1981). *Drug abuse: A guide for the primary care physician.* Chicago: American Medical Association.

Chapter 26

More Directive   Less Directive

Grief Counseling

*Losses, not fully mourned, shadow our lives, sap our energy and impair our ability to con-
nect. If we are unable to mourn, we stay in the thrall of old issues, dreams, and relationships, out of step with the present because we are still dancing to tunes from the past.*

Volkan and Zintl, 1993

HISTORY

Modern crisis intervention and, more specifically, grief counseling, began with the
work of Eric Lindmann and his associates after the devastating Coconut Grove
nightclub fire on November 28, 1942 (Zisook, 1987). This fire was the country's
[America's] worst single building fire up to that date, and resulted in the demise of
493 people (Aguilera, 1990). Lindmann was involved in helping survivors who had
lost loved ones in the fire. As a result, his clinical report on the psychological symptoms displayed by the survivors subsequently became the cornerstone for theories on
the grief process. He believed that assistance through the stages of mourning would
offset future difficulties for the grieving (Gilliland and James, 1996).

On the basis of his research, Lindmann concluded that, besides death, there are
other inevitable events in the course of one's life that generate emotional strain and
stress and lead to either mastery of the situation or failure, with impairment of functioning (much like grief resulting from death). Lindemann recognized the following
four stages of grief (Aguilera, 1990):

Stage 1   Disturbed equilibrium

Stage 2   Grief work

Stage 3   Working through the grief

Stage 4   Restoration of equilibrium

MAIN IDEAS

Loss is one of the most devastating events anyone has to face. This applies to loss of
someone, or some*thing*, held in high regard. The process of grieving occurs not only
when individuals have lost a loved one but also when they have lost a relationship, a
job, a pet, or even a dwelling (e.g., eviction, fire). It is important to understand that

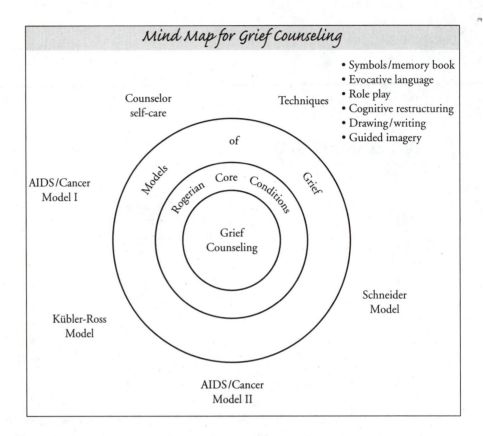

only the person suffering can categorize something as a loss. When we are able to accept the fact that we are all individuals, and every client has personal and individual paradigms through which they interpret loss, then we can provide the client with empathy and the core conditions of support (see Chapter 8). Consider the following points:

- William Worden (1991) asks the rhetorical question: "Why should mental health professionals be interested and involved in the area of bereavement?" In response, Worden cites two reasons. First, the mental health practitioner must understand and recognize the role grief plays in both medical and psychological problems. Second, grief can increase not only physical but also psychiatric morbidity.

- Gilliland and James (1996) state there are many people who are not prepared, or who are poorly equipped, to deal with grief, even though it has been part of life since the beginning of mankind.

- It is extremely important to remember or to know that grief is a *process*—an active process that necessitates experiencing painful emotions. The fast pace of modern life avails little opportunity for the slow processing of grief. Often the full impact of a loss is not felt until months later. Therefore, grief is a process with no time limit; it does not follow a straight line but rather has ups and downs. Being aware of what occurs throughout the grief process is important not only in assisting others through their grief but also in dealing with our own grief (Gallant, 1995).

- Fuller (1992) states "grief is the normal reaction to loss." It is important to remember that all losses may result in grief to some extent. These losses can span a wide range of events: death, job, relationship, fire, theft, body image, to name a few. Fuller points out that grief is far more than just sadness; in fact, it includes a range of emotions such as anger, guilt, sadness, relief, fear, and confusion.

- Worden (1991) asserts that grief can affect our whole life (work, family, friends), as well as affecting ourself psychologically, physically, and socially. Our response to various losses is dependent on whether we have had forewarning, what meaning the loss has to us, and the number of previous losses recently experienced (Fuller, 1992).

- Worden (1991) observed four specific goals in grief counseling:
To increase the reality of the loss
To help the client deal with both expressed and latent effect
To help the client overcome various impediments to readjustment
To encourage the client to say an appropriate goodbye and to feel comfortable reinvesting in life

- Gallant (1995) states that all reactions to grief are normal. People who are aware of this can feel reassured that they are not "going crazy" when they are grieving. She further explains grieving is not a weakness—it is a psychological, physical, and spiritual necessity.

- Grieving people will experience many reactions to their loss as they work toward resolution. Ultimately, the bereaved will reach a point in the recovery process when the loss becomes integrated into life experiences, and the grieving person is better able to carry on with daily living (Schneider, 1984).

- Worden (1991) describes the overall goal of grief counseling as "helping the survivor complete any unfinished business with the deceased, and to be able to say a final goodbye."

- Rando (1984) challenges professionals to come to grips with their own attitudes toward death, grief, and bereavement. Rando encourages us to look at our own history of losses in order to: better understand the process of mourning, help us get a clear sense of what resources are available, determine what worked and what didn't, and decide what to say and what not to say.

- Gilliland and James (1993) point out that for those who regularly work with loss-related clients, it is important to be involved in debriefing, have regular access to empathic listening, be aware of bereavement overload or burnout, and obtain emotional reinforcement from others (friends, family, peers).

- Assisting a grieving individual can place stress on the counselor and you need to take care of yourself while attending to the needs of the bereaved person. It is important for the counselor to maintain an adequate nutrition, sleep, and exercise program (Rando, 1984).

The two grief models we have chosen to use are the Kübler-Ross model (1969), which is the best known; and the Schneider model (1984), which is the most comprehensive. Each addresses the ways in which humans respond to loss.

The Kübler-Ross model has five stages:

Stage 1 *Denial.* Initial denial may be a healthy way of coping with the painful news.

Stage 2 *Anger.* This is a desperate attempt to gain attention and demand respect and understanding in order to establish some measure of control.

Stage 3 *Bargaining.* This is a normal attempt to postpone death.

Stage 4 *Depression.* The client is confronted with many losses that have been encountered.

Stage 5 *Acceptance.* This stage is characterized by quiet, peaceful resignation. (Kübler-Ross, 1969)

These stages provide a guide to understanding the phases of death and grieving, but they are not absolute. Not everyone will go through every stage, or in the exact sequence, or even at the same pace (Kübler-Ross, 1975).

The Schneider model has eight stages:

Stage 1 *The initial awareness of loss.* A significant stressor causes a threat to the body's sense of homeostasis.

Stage 2 *Attempts at limiting awareness by holding on.* The person uses past coping mechanisms to limit feelings of hopelessness and despair.

Stage 3 *Attempts at limiting awareness by letting go.* The person recognizes personal limits with regard to the loss.

Stage 4 *Awareness of the extent of the loss.* This stage is most readily recognized as mourning.

Stage 5 *Gaining a perspective on the loss.* Here the person accepts the loss.

Stage 6 *Resolving the loss*—time of saying good-bye, forgiveness, finishing business

Stage 7 *Reformulating loss in a context of growth.* This is an outgrowth of resolving grief.

Stage 8 *Transforming loss into new levels of attachment.* In this stage there is reintegration of the physical, emotional, cognitive, behavioral, and spiritual aspects of the person. (Schneider, 1984)

The Schneider model is also known as "the process of grieving." It is considered to be a holistic model because it integrates the person's physical, cognitive, emotional, behavioral, and spiritual responses to a loss (Gilliland and James, 1996).

Traditional models of grieving are not so effective when dealing with AIDS and cancer, where general similarities (remissions and exacerbations, unpredictability of disease course) have led to models having been specially developed for these populations. Counselors are often called upon to help clients through this tumultuous period in their lives. With this in mind, we present two models (Wilson-Nolan, 1991) to use with AIDS and cancer clients. But first we note that clients should be advised to seek legal counsel and put their affairs in order before any potential dementia occurs that could provide a loophole in executing their wishes (Greig, 1987).

Wilson-Nolan's Model 1

Stage I *Existential plight,* including profound emotional turmoil and stress

Stage II *Accommodation and mitigation,* including changes in lifestyle and a settling down of the psychosocial stressors that accompanied the initial onset of the disease

Stage III *Recurrence and relapse* typically follows a worsening of the condition and a shift from believing that a cure can be found, to seeking palliative care

Stage IV *Deterioration and decline* includes the terminal phase of the illness and the psychological preparation for death

Wilson-Nolan's Model 2

Phase I *Acute crisis,* including intense anxiety and the development of psychological resources to help deal with the crisis

Phase II *Chronic living-dying,* attempting to live a normal life despite the unpredictable fluctuations in the disease

Phase III *Terminal phase,* includes decreased anxiety and increased withdrawal

Wilson-Nolan teaches that one of the main challenges for the counselor is to move in and out of the feelings and responses of the client. Anyone facing death must adapt to a painful reality, but much can be done to make it more than a time for grieving. There can be dignity and peace in dying (Wilson-Nolan, 1991).

TECHNIQUES

There are numerous techniques that can be utilized to help a grieving person through the process. Your own eclectic approach will help to facilitate this process. Rando (1984) considers empathic listening to be the most important and useful skill that a counselor can employ with clients who are dealing with a loss.

Consider the following techniques, which we believe to be useful when helping a client work through the stages of grief:

- *Evocative language* (Worden, 1991). This consists of the counselor using tough words in order to evoke feelings (e.g., "Your husband died" instead of "You lost your husband"; the deceased is spoken of in the past tense "Your son was . . .") This is done to help the client realize that the loss is real.

- *Role playing* (see Chapter 18)

- *Cognitive restructuring* (see Chapters 16 and 18)

- *Drawing and/or writing.* This may facilitate the bereaved in expressing thoughts, feelings, and experiences in relation to the loss (Worden, 1991).

- *Guided or directed imagery* (see Chapter 18)
- *Symbols and a memory book.* The symbols can consist of pictures, letters, tapes, clothing, or jewelry that belonged to the deceased. The memory book is constructed by the family (Worden, 1991).

Summary

This chapter looked briefly at the history of grief, at the process of grief as including various models, and at a few of the numerous techniques that can be used to facilitate the process of grieving. Grief can vary widely, not only from person to person but also within the same person. Based on this premise, several different models and suggested interventions were presented. Because grief has many faces, no one can answer all your questions about grief and mourning, or tell you how to deal with someone who is going through, or stuck in, the process of grief and mourning. Skills and knowledge are the key to handling each case successfully.

References

Aguilera, D. (1990). *Crisis intervention: Theory and methodology.* St. Louis: Mosby–Yearbook.

Corey, G. (1996). *Theory and practice of counseling and psychotherapy* (5th ed.). Pacific Grove, CA: Brooks/Cole.

Fuller, J. (1992). *The community hospice handbook.* Baddeck, Nova Scotia.

Gallant, M. (1995). *Compiles from a workshop on grief.* Halifax, NS: IWK Hospital.

Gilliland, B., James, R. (1993). *Crisis intervention strategies.* Pacific Grove, CA: Brooks/Cole.

Greig, J. (1987). *AIDS: What every responsible Canadian should know.* Ottawa: Summerville Press and the Canadian Public Health Association.

Howatt, W. (1995) *Counselling for paraprofessionals: Formulating your eclectic approach.* Middleton, NS: Nova Scotia Community College.

Kübler-Ross, E. (1969). *On death and dying.* New York: Simon & Schuster.

Kübler-Ross, E. (1975). *Death: The final stage of growth.* Champaign, IL: Research Press.

Rando, T. (1984). *Grief, dying and death.* Champaign, IL: Research Press.

Schneider, J. (1984). *Stress, loss and grief: Understanding their origins and growth potential.* Baltimore: University Park Press.

Volkan, V., Zintl, E. (1993). *Life after loss.* Toronto: Macmillan.

Wilson-Nolan, B. (1991). *Home support Canada: Providing home-centred care to people with AIDS.* Ottawa: Federal Centre for AIDS, Health Protection Branch, Health and Welfare Canada.

Worden, J. W. (1991). *Grief counseling and grief therapy.* New York: Springer.

Zisook, S. (1987). *Biopsychosocial aspects of bereavement.* Washington, DC: American Psychiatric Press.

Chapter 27

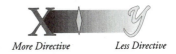

More Directive   Less Directive

Eating Disorders

I don't want to be like this, but what I eat still rules my life . . . I dream of the perfect day when I have no appetite, no thought, no desire, [no] temptation for food or to eat. I often despair of ever finding a solution.

Abraham and Llewellyn-Jones, 1992

HISTORY

Reports of self-starvation date as far back as the thirteenth century. Then it was known as "holy anorexia," and was due to a compulsive drive to seek spiritual perfection (Somer, 1995). Reports of anorexia resurfaced in the 1800s, and it became a recognized social problem during the 1920s. Today, this disorder has been related to the image portrayed by models (e.g., extremely thin, beautiful). Adolescent females account for approximately ninety percent of all cases of anorexia and bulimia (Dworetzky and Davis, 1989) It is estimated that in the United States, 1 in every 100 adolescent girls between 11 and 18 falls victim to anorexia nervosa (Somer, 1995).

Bulimia nervosa, which is characterized by binge eating, first appeared in the 1970s; it has increased so rapidly in the past 20 years that it is believed to affect 3 in every 100 girls. It is considered to be the most common eating disorder (Somer, 1995) and has been estimated to affect approximately five percent of the population (Rathus, 1990). Bulimia is about four times as common as anorexia (Dworetzky and Davis, 1989). Binge eating usually starts between the ages of 15 and 24, and follows a period of increased concern about body weight and appearance, during which the individual decides to diet or at least "watch" their weight (Abraham and Llewellyn-Jones, 1992). This disorder may be more difficult to detect because the individual may appear to be of normal size and weight (Dworetzky and Davis 1989).

MAIN IDEAS

It is not known why some people progress toward self-starvation or binge-and-purge habits (Somer, 1995). In the field of eating disorders it has been found that an adherence to strict gender stereotypes, as well as psychological issues involving personality disorders or identity and role confusion, may play a role in the development of eating

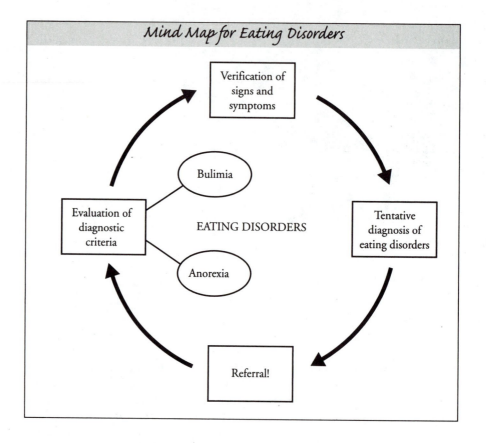

disorders. Poor reasoning skills, poor coping skills, a general body image dissatisfaction, and striving for perfection are common traits among people with eating disorders (Lenihan and Sanders, 1984). Bulimia has become a great concern on college campuses; about half of college women admit to at least one occasional cycle of bingeing and purging (Rathus, 1990). It is thought that the college atmosphere may indirectly contribute to the development and maintenance of eating disorders by emphasizing perfection, competition, and attractiveness (Kashbeck, Walsh, and Crowl, 1994). Trying to become thin may be one way for these individuals to have control and increase their self-worth (Dworetzky and Davis, 1989).

Goodman (1996) suggests that people with eating disorders often start by dieting, but then find that they cannot go without the food that calmed their hunger and emotional pangs. The next step is often bulimia (bingeing/purging), which in some cases spirals into further control of eating habits or behaviors, resulting in anorexia (Somer, 1995).

Eating disorders are the result of dysfunctional relationships with food (Brouwers, 1994). They are "characterized by gross disturbances in patterns of eating" (Rathus, 1990). This chapter will deal with two of the most common eating disorders: anorexia nervosa and bulimia nervosa. Both of these disorders involve a compulsive pursuit of thinness, often to the point of endangering the person's life (Dworetzky and Davis, 1989).

Anorexia nervosa is characterized by a prolonged refusal to eat, resulting in severe weight loss. The person often stops eating to the point of starvation (Abraham and Llewellyn, 1992). Anorexia nervosa is divided into two categories: the *restrictive type,* which relies on self-starvation for immediate weight loss, and the *bulimic type,* which involves fluctuations between bingeing and starving (Somer, 1995). Both restrictive and bulimic anorexic types will face the same long-term consequences (Somer, 1995).

People who suffer from bulimia alternate between self-starvation and excessive eating (Kalat, 1990). *Bulimia nervosa* is defined as binge eating followed by laxatives or self-induced vomiting (also referred to as bingeing and purging) (Abraham and Llewellyn, 1992).

It is important that counselors be aware of the warning signs of eating disorders so as to avoid overlooking a situation that could be life threatening. This chapter presents signs and symptoms as well as some interventions that may be utilized with this population of clients. It is also designed to assist you in screening so that clients with eating disorders may be referred to the appropriate services.

CLINICAL ASPECTS

These behaviors are classified as a disorder because they result in an imbalance of appetite-control chemicals (Rathus, 1990). Even though "dieting" may begin as a psychological issue (to lose weight), physiological factors soon play a role. Hormones and nerve chemicals in the body are disrupted as a result of erratic eating behaviors (Somer, 1995). Rathus (1990) provides some information on physiological aspects of these disorders:

- Serotonin (a neurotransmitter that is related to our feelings of well-being) levels may be directly affected by binge eating.
- Low serotonin levels stimulate hunger and are linked to depression.
- Binge eaters have a lack of serotonin that disrupts the mechanism for turning off desire for certain foods (e.g., carbohydrates).
- It is speculated that binge eating is done at an unconscious level to increase serotonin levels.
- High levels of serotonin increase feelings of satiety and calmness while allowing reduced food intake.

Somer (1995) contributes further:

- Zinc deficiency has been connected to eating disorders because it is thought to cause a loss of desire for food, poor appetite, and dissatisfaction with the taste of food.
- It has been noted that individuals with eating disorders typically consume one-half to two-thirds of the recommended daily intake of zinc (12 to 15 mg).
- Zinc deficiency may not contribute to the onset of an eating disorder but may help to maintain the problem.

Signs and Symptoms

It is essential to know the signs and symptoms of eating disorders, because the longer the disorder continues the harder it is to secure treatment and recovery (Somer, 1995).

Anorexia Nervosa

Signs and symptoms noted by Abraham and Llewllyn-Jones (1992) include:

- Brittle hair (due to lack of nutrients)
- Emaciation (skeletal, gaunt)
- Hypertension, hypothermia, and dryness of skin (due to lack of nutrients) (Ibid.)
- Lanugo (downy hair on the face, back, and arms)
- Sensitivity to hot and cold temperatures (due to lack of nutrients and loss of insulating layers of fat)
- Hands and feet may swell (retention of water following bingeing)
- Dental enamel erosion (tooth decay caused by excessive amount of acid in vomitus)
- Scars or calluses on back of hands (caused by self-induced vomiting)

Other authors add the following:

- Constant weighing of food and self (Somer, 1995)
- Amenhorrhea (cessation or absence of menses for more than 3 months) (Rathus, 1990)
- Distorted self-image (the person wrongly believes self to be fat) (Ibid.)
- Delayed puberty (body curves and sexual urges absent) (Craig, 1992).
- Deliberate undereating is believed to be responsible for about seven percent of cases of delayed puberty in the United States (Dworetzky and Davis, 1989)
- Avoidance of any "fattening" foods (Ibid.)
- Hyperactivity (Ibid.)
- Depression or irritability (Ibid.)

Bulimia Nervosa

Abraham and Llewellyn-Jones (1992) list these signs and symptoms:

- Loss of dental enamel
- Salivary glands may become enlarged, causing face and cheeks to appear puffy
- Hands and feet may swell due to the retention of fluid after binge-eating
- Calluses or scars on the back of hands (caused by self-induced vomiting)
- Exhaustion (groggy, dark circles under the eyes)

- Headaches (caused by lack of nutrition and self-induced vomiting)
- Complaints of constipation, abdominal pain, cold intolerance, and lethargy, and/or excessive energy (caused by self-induced vomiting)

Other authors add the following:

- Frequent vomiting (average twice a week for three months) (Somer, 1995)
- Secretive behavior (bingeing/purging while no one is home, wearing oversize clothes, hiding food)
- Rapid consumption of food in a short period of time (gorging) (Craig, 1992)
- Depression or irritability (Dworetzky and Davis, 1989)
- Guilt or shame about eating (Ibid.)

Diagnostic Criteria for Eating Disorders (DSM-IV)

We have included these criteria so that the human services counselor can check personal observations and assumptions against a reputable source. Human services counselors are not permitted to diagnose, but we believe it is important for you to be familiar with the most frequently used diagnostic tools as you work with other professionals in the counseling field. The *Diagnostic and Statistical Manual of Mental Disorders, Fourth Edition (DSM-IV)* (1994) has undisputed credibility as the most respected reference tool in the field today.

Anorexia Nervosa

A. Refusal to maintain body weight at or above a minimally normal weight for age and height (e.g., weight loss leading to maintenance of body weight less than 85% of that expected; or failure to make expected weight gain during period of growth, leading to body weight less than 85% of that expected).

B. Intense fear of gaining weight or becoming fat, even though underweight.

C. Disturbance in the way in which one's body weight or shape is experienced, undue influence of body weight or shape on self-evaluation, or denial of the seriousness of the current low body weight.

D. In postmenarchal females, amenorrhea, i.e., the absence of at least three menstrual cycles. (A woman is considered to have amenorrhea if her periods occur only following hormone, e.g., estrogen, administration).

Specify type:

Restricting type: During the current episode of Anorexia Nervosa, the person has not regularly engaged in binge-eating or purging behavior (e.g., self-induced vomiting or the misuse of laxatives, diuretics, or enemas).

Binge-eating/Purging Type: During the current episode of Anorexia Nervosa, the person has regularly engaged in binge-eating or purging behavior (i.e., self-induced vomiting or the misuse of laxatives, diuretics, or enemas). (DSM-IV, 1994)

Bulimia Nervosa

A. Recurrent episodes of binge-eating. An episode of binge-eating is characterized by both the following:

(1) Eating, in a discrete period of time (e.g., within any 2-hour period) an amount of food that is definitely larger than most people would eat during a similar period of time and under similar circumstances.

(2) A sense of lack of control over eating during the episode (e.g. a feeling that one cannot stop eating or control what or how much one is eating).

B. Recurrent inappropriate compensatory behavior in order to prevent weight gain, such as self-induced vomiting; misuse of laxatives, diuretics, enemas, or other medications; fasting; or excessive exercise.

C. The binge-eating and inappropriate compensatory behaviors both occur, on average, at least twice a week for three months.

D. Self-evaluation is unduly influenced by body shape and weight.

E. The disturbance does not occur exclusively during episodes of Anorexia Nervosa.

Specify Type:

Purging Type: During the current episode of Bulimia Nervosa, the person has regularly engaged in self-induced vomiting or the misuse of laxatives, diuretics, or enemas.

Non-purging type: During the current episode of Bulimia Nervosa, the person has used other inappropriate compensatory behaviors, such as fasting or excessive exercise, but has not regularly engaged in self-induced vomiting or the misuse of laxatives, diuretics, or enemas. (DSM-IV, 1994)

INTERVENTIONS

An awareness of the goals of interventions provides the counselor with the knowledge to make an informed referral to the medical or psychological professional, which can be essential for the survival of the client. Abraham and Llewellyn (1992) cite the goals as follow:

Anorexia Nervosa

1. Increase body weight slowly (2 lb per week)
2. Stop the use of dangerous weight loss methods
3. Learn sensible eating habits
4. Decrease preoccupation with weight and food

Bulimia Nervosa

1. Stop trying to lose weight
2. Stop weight loss behavior

3. Learn sensible, normal eating patterns (at least three balanced meals a day)

4. Decrease preoccupation with weight and food

5. Stabilize weight in a desirable range

As to actual interventions with this population, the human services counselor will always choose referral when this life-threatening situation arises. From my research, one of the most successful clinics in the world is the Montreaux Clinic in Vancouver, BC, operated by Peggy Claude Pierre. If the counselor is working as part of a multidisciplinary team, however, we suggest using the individual eclectic orientation in a manner congruent with the multidisciplinary team's plan of action and always acting in the client's best interest. Stewart (1994) suggests that assertiveness training (see Chapter 18) is a highly effective tool in the treatment of eating disorders. I believe that a biopsychosocial approach provides an effective holistic foundation for working with this population. Once individuals have been treated for their eating disorder, they may still require individual counseling. The human services counselor can then once more play an effective role.

Summary

The problems of anorexia and bulimia have existed for hundreds of years, and it appears they will continue. Professionals encountering individuals who have an eating disorder need to provide support, empathy, and options that will help to empower the client in overcoming the eating disorder. It is essential to be knowledgeable about the signs and symptoms and diagnostic criteria. I strongly suggest that professionals be aware of their level of competency. It is important that you consult with your supervisor and provide an appropriate referral when you perceive your client may meet the criteria for an eating disorder, because you are grappling with a life-threatening disorder.

References

Abraham, S., Llewellyn-Jones, D. (1992). *Eating disorders: The facts.* New York: Oxford University Press.

American Psychiatric Association. (1994). Diagnostic and statistical manual of mental disorders (4th ed.) Washington, DC: Author.

Brouwers, M. (1990). Treatment of body image dissatisfaction among women with bulimia nervosa. *Journal of Counseling and Development 69:*144–47.

Buckroyd, J. (1996). *Anorexia and bulimia: Your questions answered.* Rockport, MA: Element Books Limited.

Craig, G. J. (1992). *Human development* (6th ed.). Englewoods Cliffs, NJ: Prentice-Hall.

Dworetzky, J., Davis, N. (1989). *Human development: A lifespan approach.* St. Paul: West.

Goodman, E. (1996, January). *The pressure to be perfect. Glamour:* 154ff.

Kalat, J. W. (1990). *Introduction to psychology* (2nd ed.) Belmont, CA: Wadsworth.

Lenihan, G., Sanders, C. (1984) Guidelines for group therapy with eating disorder victims. *Journal of Counseling and Development 63:*252–54.

MacDonald, D. (1996). Distorted images: Anerexia nervosa and bulimia nervosa. *Canadian Journal of Continuing Medical Education 8:*43–59.

Pipher, M. (1995). *Hunger pains: The modern woman's tragic quest for thinness.* New York: Ballentine.

Pipher, M. (1994). Reviving Ophelia: Saving the selves of adolescent girls. New York: Putnam.

Rathus, S. A. (1990). *Psychology* (4th ed.). New York: Holt, Rinehart, and Winston.

Somer, E. (1995). *Food and mood: The complete guide to eating well and feeling your best.* New York: Henry Holt.

Stewart, L. (1994). *Anorexia nervosa.* Unpublished master's thesis. Wolfville, NS: Acadia University.

Chapter 28

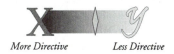

Understanding Stress

Optimal mental [and physical] health requires the acquisition of skills and characteristics to effectively encounter and grow through setbacks, opportunities, challenges, adversity, stressors, and other life events. The experience of growing through adversity is resiliency.

Bruess and Richardson, 1995

HISTORY

The human body has been equipped with an automatic coping mechanism that is activated when a threat is present. Hans Selye, the famous Canadian researcher, named this the *fight or flight* response. This response enables the body to immediately mobilize all of its resources for quick action. Bruess and Richardson (1995) list the following indicators as being part of the fight or flight response:

1. Sharp rise in blood pressure

2. Increase in blood sugar

3. Quick conversion of glycogens (stored carbohydrates) and fats into energy so as to sustain high energy utilization

4. Increased respiration to increase the body's supply of oxygen

5. Increased muscle tension, which allows for quick applications of strength

6. Pupil dilation for visual acuity

7. A release of *thrombin,* which is a blood clotting hormone to prevent excessive bleeding from potential wounds

This coping mechanism is one of the reasons humanity has evolved. In primitive times, it allowed for a better chance of survival against a predatory animal, and in civilized societies it still serves to protect the individual.

Hans Selye (1976) was a pioneer in the field of stress. He studied how stress affects the body. His work helps us understand how overstimulation and excessive wear-and-tear on the mind and body may lead to stress-related dysfunction and disease. While *stress* is defined as the nonspecific response of the body to any demand made upon it, Selye described the stress response as a three-stage process called the *general adaptation syndrome* (Selye, 1976). Preceding the onset of the first stage, there

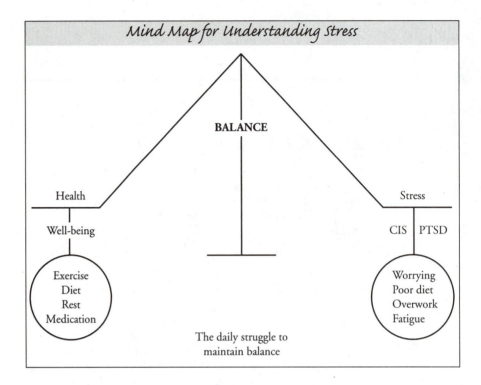

may be a conscious or unconscious evaluation in which a decision is made about whether a threat exists.

Williams and Knight (1994) presented the three stages of Selye's general adaptation syndrome (Figure 28–1) as follows:

1. The *alarm phase* (fight-or-flight response) is the direct effect of the stressor upon the body. Muscles tense, respiration rate increases, blood pressure and heart rate increase.

2. The *resistance phase* involves the body's adaptation to the challenge of the stressor. The body's level of strength and endurance increases, if necessary, or returns to a level of normal activity. During this phase, the individual begins to concentrate more on psychological coping mechanisms and defensive behavior to deal with the stressor, rather than continuing the physical fight or flight response.

3. The *exhaustion phase* involves the body's collapse to the stress and the stressor. The body's physical and psychological energy has become depleted. The individual becomes exhausted and, without rest, may become ill, or in extreme cases may die.

Understanding the dynamics of Selye's general adaptation syndrome may help you arrest the stressing process by taking better care of yourself. This is particularly important for those in the helping professions because we are prone to burnout.

Figure 28–1.
Selye's general adaptation syndrome (adapted from Selye, 1976)

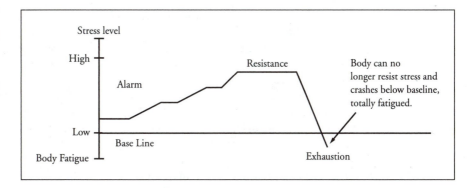

Burnout is described by Williams and Knight (1994) as a state of physical, emotional, and mental exhaustion, as well as a syndrome of emotional exhaustion, depersonalization of others, and a feeling of reduced personal accomplishment. Burnout is generally brought about by job-related stress, although it is also common in family caregivers. We may recognize this state in our colleagues, in ourselves, or in our clients, who have simply become overwhelmed by the pressures of life with little support.

MAIN IDEAS

Human services counselors, by nature of their vocation, extend themselves to others and offer support for growth and development on a daily basis. Gilliland and James (1996) state the work environments of helping professionals often carry with them harsh realities that may make workers prone to stress. Along with personal demands, workers often have to contend with long and erratic hours, unpredictability as to their clients' functioning, large caseloads, and budget constraints. Clients and colleagues alike find themselves affected by stress, which may have negative consequences for personal and professional productivity, and for physical and mental health. Paradoxically, the most important thing the human services counselor can do for clients and workplace is to be diligent in the practice of self-care.

Stress has become a buzzword of the nineties that has a decidedly negative connotation. But stress can have both positive and negative effects on one's life. In fact, we humans need a certain amount of stress to motivate us. Hanson (1986) has suggested that *optimal* stress loads result in longevity, peak performances, and the balance between adequate stimuli and overloading. Indeed, if people find they have nonproductive time, they should say yes more often to challenges (Hanson, 1986).

Achieving an optimal stress level involves understanding the difference between good stress *(eustress)* and bad stress *(distress)*. Good stress may result from such things as falling in love, buying a house, or getting a new job. If you experience good stress you are likely to feel more focused, more energized, more motivated, more aware of your options—challenged by the present opportunity. Bad stress may result from an event or situation that taxes you in some way. If you experience bad stress, you are likely to feel tense, anxious, angry, depressed, and frustrated. Brewer (1991) referred to a chronic, pervasive, subtle form of bad stress as "ugly" stress. This type of stress is typically present when you feel exploited over an extended period of time, are uncertain about your future, or suffer financial difficulties. Brewer suggests this type of stress can be the most severe form and causes lack of energy, chronic depression, health problems, and low self-esteem.

It is important to note that stress is a highly individual experience. What is stressful for one person may be exhilarating for another. For example, one person may get a thrill out of speaking in front of a large audience, but to someone else public speaking may be a terrifying thought. To understand what sort of things negatively affect you, it may be helpful to examine the different sources of stress. A *stressor* is a source of stress, which may come in the form of "external cues (people, situations, elements), internal mental processes (worry, fear, happiness), or physiological processes (drugs, sugar, biorhythms) . . ." (Bruess and Richardson, 1995).

Gilliland and James (1996) identify two types of stressors: psychosocial and biogenic. *Psychosocial stressors* are such because of the cognitive interpretation assigned to the stimulus; that is, they are stressors because we perceive them as such (a traffic jam). *Biogenic stressors* (caffeine, alcohol, drugs, exercise) are such because they possess some electrical or biochemical property capable of stimulating the stress response.

An enormous variety of life events have the potential to create stress, ranging from daily hassles (housework, traffic, paying bills) to life changes (marriage, death of a loved one, change in job status). Williams and Knight (1994) teach us that both magnitude and frequency of life changes produce stress in an individual's life. If you have experienced a lot of changes in your life recently, likely you are more vulnerable to the adverse effects of stress. As well, if you are the type of person who is always in a hurry, has many things on the go at one time, and is impatient and competitive, you are also likely to be more prone to stress.

Brewer (1991) states that if you see yourself as a victim, doubt your ability to control your future, and don't see uncertainty as a potential opportunity, you may rarely experience good stress. Conversely, if you are optimistic about the future, feel in control of your life, and see yourself as having the stamina and internal resources to bounce back from adversity, you will suffer ugly stress infrequently and, when it does occur, you will probably recover quickly. In other words, you can reduce your level of stress by developing a more positive mental attitude (PMA) and by viewing situations as challenges rather than setbacks. We should view the world in the same way as the famous umpire who said "It ain't nothing until I can call it something!"

CRITICAL INCIDENT STRESS (CIS)

A key to managing stress is to recognize the signs and symptoms related to it. Doctors estimate that 75% of all medical complaints are stress related; 50% of the population suffers from at least one stress-related symptom on a regular basis (Brewer, 1991). Table 28–1 summarizes some physical, psychological, and behavioral signs that may be attributed to stress. If you suffer from any of these symptoms regularly it is important to consult with your physician. This may seem obvious, but we all

Table 28-1. Signs and symptoms of stress.

Physical	Psychological	Behavioral
Insomnia or fatigue	Anger	Biting lips
Sexual dysfunction	Anxiety	Drug and/or alcohol abuse
Indigestion/nausea	Apathy	Foot tapping or turning
Constipation	Boredom	Grinding teeth
Ulcers/diarrhea	Depression	Impulsive actions
Headaches	Fatigue	Increased smoking
Muscle aches or spasms	Fear of death	Isolating from family and friends
High blood pressure	Frustration	Moving in tense, jerky ways
Dizziness or fainting	Guilt	Nervous tics
Chronic illness / flu or colds	Hopelessness	Trembling hands
Back pain	Hostility	Overreacting
Excessive perspiration	Impatience	Rapid mood swings
Over/under eating	Inability to concentrate	Stuttering
Pounding heart	Irritability	Swearing
Shortness of breath	Restlessness	Touching hair, ears, or nose
Skin rashes		Child/spouse abuse
Dry mouth		

(Adapted from Brewer, 1991, and Spence, 1988)

know there are times when we self-diagnose and then try to self-medicate our symptoms.

These are all normal reactions to everyday stress. But sometimes there are critical situations in life that are outside the range of normal coping abilities. These events can trigger what is termed *critical incident stress (CIS)* that, in turn, can put the individual at risk for developing what is called *posttraumatic stress disorder (PTSD)*. The psychological reaction known as PTSD will be discussed more fully at the end of this chapter.

Critical incident stress is emotional trauma—a blow to the psyche. A *critical incident* occurs when individuals are exposed to external stimuli that they perceive to be outside the realm of normal (seeing, or hearing about, a catastrophic event) (Mitchell, 1988). Whenever individuals are exposed to critical incidents, there is the possibility that they will develop symptoms similar to PTSD; thus, they must be taken very seriously. Everly (1992) listed the early warning signs of PTSD as follows:

- Bad dreams, sleep disturbances
- Increased depression

- Increased self-medication (substance abuse)
- Explosive behaviors
- Changes in eating habits
- Emotional outbursts
- Irritability/anger
- Withdrawal
- Change in activity
- Signs (what we see) and symptoms (what they tell us) may occur at the scene, within hours, days, weeks, or months

Everly first proposed the term *psychotraumatology* in February 1992 at a lecture in Montreux, Switzerland. In this lecture, Everly defined the term as the study of psychological trauma caused from a critical incident. He added that the term involves the study of relevant psychological processes both before and after a traumatic event. Everly believes the term is broader than PTSD because PTSD only refers to a psychological process after the event, whereas psychotraumatology looks at the individual through a biopsychosocial window.

If the client continues to manifest early warning signs for more than 30 days, odds are increased that the person may go on to develop PTSD. Mitchell (1988) teaches that all of these symptoms are a normal response that originates from an unnormal situation. If symptoms are addressed early, the success rate is much greater, and both financial and professional costs are reduced.

Research has shown that, once an individual has developed PTSD, it is extremely time-consuming and expensive to treat (Mitchell, 1992). The rationale for *critical incident stress intervention* thus is to intervene with the individual before the symptoms become engrained.

The process of recognizing, preventing, and treating CIS was integrated through the work of Jeffrey T. Mitchell of Maryland (Mitchell, 1987). Mitchell (1992) points to four major influences that contributed to the CIS process: (a) military experience, (b) police psychology, (c) emergency medical services, and (d) disasters. In 1983, the first article discussing CIS was published in the *Journal of Emergency Medical Services*. This article was the culmination of nine years of developing the CIS process. Since that publication, emergency service organizations across the globe have adapted their strategies for dealing with traumatic events to include *critical incident stress debriefing (CISD)*. For more information, contact the International Critical Incident Stress Foundation at (410) 730-4311.

Mitchell (1988) explains critical incidents as events that are outside the realm of normal human activity or experience for the individual involved, and are so overwhelming that normal coping mechanisms do not suffice. These events can be highly stressful and fast-moving, requiring quick decisions to be made. The CIS reaction may occur immediately, or even days later, depending on the nature of the event and the individual involved. It is essential to realize that even the strongest of wills can be affected by advanced stress (Mitchell, 1987), and to note that not everyone is affected by CIS in the same way.

Everly (1992) lists a number of warning signs or symptoms of CIS:

- *Physiological*—sweating, chills, nausea, fatigue, muscle tremors, shakiness, and lack of coordination
- *Cognitive*—memory loss, decreased ability to solve problems or make decisions, concentration problems, or loss of sense of time
- *Emotional*—powerful expressions of anxiety, fear, guilt, frustration, depression, detachment, and feeling overwhelmed
- *Behavioral*—changes in usual behavior such as eating habits, sexual drive, alcohol or tobacco use, sleeping habits, and even social routine

Everly (1992) goes on to provide suggestions for things to be done or avoided in the *four days following* the exposure to a critical incident so as to reduce the chances of posttraumatic difficulties:

- Do a defusing or formal debriefing, whichever will best meet the need
- Reduce caffeine and sugar intake
- Avoid alcohol completely
- If the event involved a group of people (natural disaster, suicide of a peer, fire) encourage peers to ask in a non-threatening manner "Are you OK?"
- Do not replay events and self-blame: you did the best you could in a bad situation
- Eat lots of fruit and vegetables, little meat
- Increase rest
- Exercise briskly each day
- Remember that attitude is important, laughter reduces stress
- Remember your reactions are normal
- Don't fight sleeplessness—get up and read, walk, or write
- Do things that feel good
- Be aware of effects on family and friends

In a lecture, Mitchell (1992) recommended using educational interventions in dealing with CIS. He noted that the effects of CIS can be greatly reduced by using the techniques of *defusing, demobilization,* and *debriefing.* These techniques are not therapies, they are educational strategies aimed at getting affected individuals back to work and into their normal routine as soon as possible. **Important note:** These techniques must only be conducted by individuals trained in CISD. The underlying process of CIS is much more complicated than it may appear on the surface.

Defusing is a shortened version of debriefing, usually lasting 20 to 45 minutes (as opposed to 2 or 3 hours for debriefing). It is generally in the form of a small-group meeting provided as soon as possible (no more than 8 hours) after the conclusion of a traumatic event. Defusings are aimed at the core groups most seriously affected by an event. Defusing may eliminate the need for a formal debriefing. Defusings are composed of an introduction followed by an exploration and information phase, and they are designed to provide constructive ways for people to discuss and deal with their stress. There are three stages to a defusing:

1. Introduction — quick remarks (confidentiality)
2. Exploration — tell about incident
   - What is bothering you most about the incident?
   - What is the worst thing you are experiencing?
3. Information — give all the factual information about the situation
   - Straighten out facts and opinions (explore mixed messages)
   - Identify positive responses
   - Explore any past experiences
   - Discuss how much time they had to make a decision

Demobilizations are designed to deal with a large-scale incident such as a multi-casualty disaster and are a direct substitute for defusing (either one or the other will be used, but not both). The demobilization is divided into two segments. The first part is about 10 minutes long and involves passing information to large groups of people. Because of the numbers, there is little opportunity for interaction in this segment, however, in the second segment, when the group has moved to a separate room for relaxation and (maybe) food, CISD team members are available to talk. After a demobilization, participants should not be allowed to return to the scene; for example, demobilizations should only done at the end of a shift. "This is a primary prevention intervention technique to help deal with large numbers of potentially distressed personnel to keep the clients balanced until a debriefing can be organized" (Mitchell, 1987).

The main focus of *critical incident stress debriefing (CISD)* is to allow the participant to vent emotions and feelings, and to reduce the negative impact of the CIS. It also helps to bring together in an internally supportive manner the groups who respond to these critical incidents for peer support.

The two types of debriefs are the spontaneous, unplanned debriefs, and the planned, formal debriefs. *Spontaneous debriefs* include hot debriefs, done at or near the scene for workers who have become overwhelmed on the spot. The venting of emotions about how the worker is responding to the situation and facilitation, reassurance, support, and acceptance make up the hot debrief.

Planned debriefs consist of formal debriefs and referrals or follow-ups. The formal debrief should take place 24 to 72 hours after a critical event (Mitchell, 1989a). The planned debrief consists of setting rules and procedures, re-creating what happened from each individual's unique perspective—focusing on the who, what, where, and when, and then noting the ways in which participants differ on specifics. The how and why of the incident is also examined. There are seven stages (Figure 28–2) to the formal debrief (adapted from Mitchell, 1992):

1. *Introduction:* where the boundaries of debriefing are delineated
2. *Fact phase:* people tell who they are, what they saw, and what their job was
3. *Thought phase:* participants describe their thoughts and reactions in relation to their experiences with the critical incident

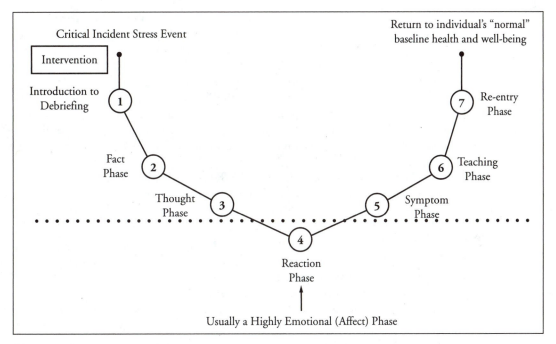

Figure 28–2.

Mitchell's seven-step debriefing model (adapted from Mitchell, 1989)

4. *Reaction phase:* individuals explain what was the worst part of the event for them; encouragement to vent feelings is provided, giving each a chance to express anger, sadness, and frustration

5. *Symptom phase:* describing the symptoms being experienced, such as shock and numbing

6. *Teaching phase:* participants are encouraged to ask questions about reactions they are experiencing, so they can have a better understanding of what is happening, and what can be done

7. *Re-entry phase:* plans are made to help participants function normally again with minimal residual effects; also included is establishment of contacts for referrals and follow-ups for individuals who desire them; additional assistance by professional CISD trainees

Again we note that these techniques must only be conducted by individuals trained in CISD.

POSTTRAUMATIC STRESS DISORDER (PTSD)

Posttraumatic stress disorder (PTSD) has probably been in existence since humans have been capable of rational thought (Gilliland and James, 1996). However, it was not viewed as a problem until World War I, when a condition ("shell shock") resulting from the traumatic war experience was recognized; in World War II the term was "war neurosis" (Everstine and Everstine, 1993). It was not until the Vietnam experience that PTSD was brought into the awareness of human services and the public.

"Psychic trauma is a process initiated by an event that confronts an individual with an acute overwhelming threat" (Gilliland and James, 1996). This threat or event is most often unexpected and experienced both "intimately and forcefully" (Everstine and Everstine, 1993). It is when this event or trauma is not effectively resolved and becomes buried in the individual's subconscious awareness that the danger of PTSD arises. In this group of individuals, the chance is increased that the initiating stressor will reemerge, masked as any number of symptoms, months or even years after the event.

The first diagnostic description of PTSD as an identifiable disorder occurred in the American Psychiatric Association *Diagnostic and Statistical Manual of Mental Disorders, Third Edition* (DSM-III) in 1980. According to the *Fourth Edition* (DSM-IV, 1994):

> . . . the essential feature of PTSD is the development of characteristic symptoms following exposure to an extreme traumatic stressor involving direct personal experience of an event that involves actual or threatened death or serious injury, or other threat to one's physical integrity; witnessing an event that involves death, injury, or threat to the physical integrity of another person; or learning about unexpected or violent death, serious harm or threat of death or injury experienced by a family member, or other close associate.

To diagnose PTSD, the criteria set out by the DSM-IV must be met. These criteria are presented in the following *only* in order to increase knowledge and awareness among human services professionals. By no means should human services counselors make PTSD diagnostic assessments. Because PTSD is one of the more elusive disorders to diagnose accurately, it requires a professional diagnostician with many years of training and experience. A diagnosis of PTSD may result in long-term treatment, perhaps involving such things as pharmacology, redecision therapy, and imprinting.

As human services counselors, it is our job to be aware of situations that put a client at risk for PTSD so we can refer these cases to the appropriate professional. Here are the DSM-IV (1994) criteria for PTSD:

I. The person will have experienced an event that is outside the realm of usual human experience, and that would be severely distressing to almost anyone. Examples are a serious threat to one's life or physical integrity; serious threat or harm to one's children, spouse or other close relatives and friends; sudden destruction of one's home or community; or seeing another person seriously injured or killed.

II. The traumatic event is persistently reexperienced in at least one of the following ways:

1. Recurrent and intrusive distressing recollections of the event.

2. Recurrent nightmares of the event.

3. Flashback episodes that may include all types of sensory hallucinations or illusions that cause the individual to dissociate from the present reality and act or feel as if the event were recurring.

Stimuli that in some way resemble the past event may precipitate the reexperiencing. Such stimuli may range from the anniversary of the event to a smell associated with it.

III. The person persistently avoids such stimuli through at least two of the following ways:

1. Attempts to avoid thoughts or feelings associated with the trauma.

2. Tries to avoid activities or situations that arouse recollections of the trauma.

3. Has an inability to recall important aspects of the trauma.

4. Has markedly diminished interest in significant activities.

5. Feels detached and removed emotionally and socially from others.

6. Has a restricted range of affect by numbing feelings.

IV. The person will have persistent symptoms of increased nervous system arousal that were not present before the trauma, as indicated by at least two of the following problems:

1. Difficulty falling or staying asleep.

2. Irritability or outbursts of anger.

3. Difficulty concentrating on tasks.

4. Constantly being on watch for real or imagined threats that have no basis in reality (hypervigilance).

5. Exaggerated startle reactions to minimal or nonthreatening stimuli.

6. Physiologic reactivity upon exposure to events that symbolize or resemble some aspect of the trauma such as a person who was in a tornado shaking violently at every approaching storm.

V. The duration of the foregoing symptoms must be at least one month. Delayed onset is specified if the start of the symptoms is at least six months after the trauma.

ALL FIVE OF THE SYMPTOM CRITERIA MUST BE MET IN ORDER FOR A PTSD DIAGNOSIS TO BE MADE.

TECHNIQUES

Having referred those clients who have CIS or PTSD, we now turn to a consideration of the ordinary stress experienced by most clients and counselors at some time in their lives. Murray (1995) states that people's ability to improve their immediate and long-term response to stress will result from their employing some type of stress management program. The following are some techniques that may help everyone improve overall physical and mental health and reduce everyday stress.

1. *Nutrition.* Energy is produced by metabolism, the processing of what we eat and drink (Lowen, 1989). Modifying our diets to achieve proper weight control

directly alters our energy level. When we are more sedentary and take in excess calories, our weight increases and our energy level decreases. Choosing a balanced diet ensures adequate intake of essential elements of nutrition, these being carbohydrates, proteins, fats, minerals, water, and fiber (Kuntzleman,1987). Werbach (1992), states "It is now well established, for example, that nutritional factors are of major importance in the pathogenesis of both atherosclerois and cancer, the two leading causes of death in Western countries, and studies validating their importance in the pathogeneses of many other diseases continue to be published."

Canada's Food Guide (1992) asserts that different people need different amounts of food. Excess amounts of fats and sugars should be avoided. The amount of food you need will depend on your age, sex, exercise level, and overall health (Howatt, 1995). It is recommended that individuals consume eight (8) glasses of water per day, particularly when under stress. Drinking appropriate amounts of water increases blood flow (one of the body's response to stress is to thicken blood for clotting), regulates body temperature, aids digestion, and assists in the elimination of toxins and waste. It is also beneficial to limit salt, fat, caffeine, and alcohol.

2. *Exercise.* Exercise provides a twofold advantage to stress relief (Brewer,1991). First, working the muscles, heart, and lungs provides direct relief of daily stress. Second, strengthening the muscles, bones, and tendons through exercise better equips us to deal with future stress and anxiety. Exercise is not only beneficial to your body but also your mind. During exercise, the individual has an opportunity to disregard personal concerns while concentrating on the activity. Exercise also provides opportunity for interaction with others, which in itself can reduce stress (Orlandi and Prue, 1988). Examples:

- Counselors can help clients work out exercise schedules that fit their needs and lifestyles (a senior who lives alone may enjoy walking with a group for an hour a day, 3 to 4 times per week).

- Counselors can suggest a variety of exercise options (walking, dancing, lawn bowling, skating, swimming).

- Counselors can suggest joining the local fitness center.

- It is suggested that, in order to reap the full benefits of stress reduction through exercise, you should engage in an enjoyable aerobic activity for at least 30 minutes at a time, at least 3 times a week (Brewer, 1991).

- Individuals can join a friend to walk, swim, bike, or go to a fitness center.

3. *Rest.* The average adult requires 6 to 9 hours of sleep a night in order to rest the body completely. We need to proceed through a complete sleep cycle in order to fully recover from a day's activities. Sleep gives the body a break for the stress of the day and allows us to prepare for the next day. A disturbed sleep can prevent us from feeling rested, so sometimes naps through the day may be needed. It is best to go to bed at approximately the same time each night and to get up at the same time each day (Hanson, 1986).

4. *Meditation.* The practice of meditation is an excellent way to relax and reduce tension, thereby reducing stress. Meditation can be learned from a personal instructor, a center offering classes, or programmed tapes or books (Brewer, 1991). Novice meditators start by focusing on their breathing. An easy way to begin is to sit in a darkened room for 20 minutes, focusing on the act of breathing. At first it is natural for the mind to wander so it can help to think of the word "in" when inhaling and "out" when exhaling (Hookam, 1988). *Shamantha* meditation (meaning resting in tranquility) focuses on the out-breath, and how it leaves the body. Body posture is upright and alert, for the person to feel a sense of dignity and presence. According to Hookham (1988), "Once the mind learns to abandon its ambition and stops preoccupying itself with endless activities, it becomes very open and relaxed and the sense of time and space, self and others and so on, begin to become transparent. It is then that true compassion can emerge. . . ."

5. *Deep breathing.* Deep breathing is a technique we can use to re-energize, relax, and regain control of our emotions. Lowen (1969) explains that deep breathing releases tensions through the body; the client is encouraged to breathe deeply into the abdomen. Take a deep breath, inhaling through the nose and exhaling through the mouth. Take progressively longer breaths and focus on the sound and feeling of the breath going in and out of the body. Continue this process for 5 to 10 minutes, and practice it once or twice per day.

6. *Stretching.* Stretching is a technique that both client and professional can use to reduce stress and tension. Stretching also increases flexibility (Howatt, 1995). One popular practice that involves stretching is yoga. Working with a yoga practitioner may also be effective at lowering the heart rate, slowing metabolism, and lowering body temperature. However, the user-friendly benefit for a counselor is the concept of stretching. Walsh (1989) suggests that yoga's strong emphasis on body posture and breath control may help the client to release body tension. Yoga training is both cost-effective and accessible to members of Western society (Walsh, 1989). You can refer to your local yellow pages for classes in yoga.

7. *Muscular relaxation.* When the counselor notices areas of muscular tension in clients, it is possible, through body positioning and breathing, to release emotional and physical pain (Lowen, 1989). Lowen (1969) believed the most effective method for muscle relaxation is through body work. In today's society, an increasingly popular way to relax is to release tension is through massage. Massage therapy relaxes muscles and helps enhance the sense of well being.

8. *Biofeedback.* Biofeedback is a technique, used in conjunction with other therapeutic techniques, in which people are trained to respond to their own bodily signals (Microsoft Encarta, 1994). They receive feedback about a reaction that is taking place in their system and can learn how to change it. It is primarily used in treating painful or stress-related conditions and works by helping a client control physiological responses that are normally considered involuntary. Biofeedback has been likened to learning an athletic skill—it takes practice. The benefits seen through the use of biofeedback typically occur because of the

relaxation that is induced by the procedure (Microsoft Encarta, 1994). The client monitors a particular stress-induced physiological reaction, such as a racing heart, and then uses relaxation techniques to lower it. (Using biofeedback requires special equipment and training.)

9. *Time management.* Time management means learning how to plan your time more effectively, not putting more activities into less time. Practicing good time management techniques at work can mean you have time to spend with family and friends, free from worry and guilt. Effectively managing your time can include the following activities:

- Buy a day planner
- Set priorities
- Organize your day
- Delegate authority; learn to train and depend on others
- Tackle tough jobs first
- Minimize meeting time; schedule meetings before lunch or at the end of the day
- Avoid putting things off
- Don't be a perfectionist; do your best in a reasonable amount of time

10. *Support systems.* Relationships can play several roles in our lives. Of these, the most important is that of personal support during times of need. We all know that talking with someone who cares about us is a very relieving experience. Thus, it is important that we maintain these nurturing relationships so they will be available when we need them. Remember, you were a person with needs of your own before you became a counselor! Invitational counseling focuses on one's own personal existence as seen by the self. To help us view ourselves in a positive manner, both personally and professionally, I offer the following suggestions adapted from *Invitational Counseling* (Purkey and Schmidt, 1996):

a) Give yourself a celebration. Make a pledge to do something special for yourself, and only yourself, in the immediate future.

b) Rehearse the future, not the past. So often when we make mistakes, we go over them again and again in our minds—in effect practicing the mistakes. A better way to overcome the mistake is to ask, "How will I handle this concern the next time?"

c) Practice positive self-talk. Negative self-statements—such as, "I could never do that"— can discourage one from perceiving oneself as able, valuable, and capable. Change negative self-talk to positive internal dialogue.

d) Find a way to exercise. Professionals can be more inviting when they maintain their own physical health.

e) Talk with a friend. A good way to prevent professional burnout is to talk with a friend whose judgment you trust.

f) Plant a garden. Working in the soil can be a relaxing and rewarding experience, particularly when you watch a plant grow and produce.

g) Build a personal Fort Knox. Start a special file of letters, awards, notes, gifts, and other recognition you have received over the years. (Stash of Strokes!)

h) Form positive food habits. When planning your meals or eating out, try meats that are broiled, baked, or roasted rather than fried or sauted.

i) Renew a friendship. Call or write and old friend you have not talked with in a long time.

j) Live with a flourish. Avoid drabness, gain satisfaction from many sources, find ways to enrich your life; stand tall, dress well, eat less, and surround yourself with things you like.

k) Live a longer life. People live longer when they take responsibility for their physical health.

l) Explore a library. For a relaxing experience, spend several hours browsing in a library.

m) Plan an adventure. Visit travel agencies and load up on travel brochures.

n) Raise your drawbridge. Although too much isolation is not good, some time alone to enjoy stillness, to contemplate and meditate on who you are, where you came from, and where you are going contributes to living, and can be both personally and professionally rewarding.

o) Take a few risks. When chances of success are good, it usually pays to take a few chances.

p) Laugh. Subscribe to a happy little magazine titled *Laughing Matters*. It is published quarterly by Dr. Joel Goodman, Director, The Humor Project, 110 Sprint St., Saratoga Springs, New York, 12866, and it will brighten almost any day.

q) Be a good physician. Healthy people work at being good physicians to themselves by taking responsibility for their own well-being.

Summary

This chapter provides insight on both critical incident stress (CIS) and posttraumatic stress disorder (PTSD). It includes ways to reduce the effects of CIS and warning signs for PTSD, which must always be referred for treatment.

A second purpose of this chapter was to provide human services counselors with the resources to keep themselves healthy and well. Once you are able to incorporate these tools into your own lifestyle, you will be able to provide your clients with a positive role model of healthy living.

My class's motto is "Live, Love, Learn, Laugh, and Have Some Fun." As a group, we truly believe health and well-being are the key to productivity and success in this challenging field of service.

References

American Psychiatric Association. (1994). *Diagnostic and statistical manual of mental disorders* (4th Ed). Washington, DC: Author.

Brewer, K. C. (1991). *The stress management handbook*. Shawnee Mission, KS: National Press Publications.

Bruess, C., Richardson, G. (1995). *Decisions for health*. Toronto: Brown and Benchmark.

Health and Welfare Canada. (1992). *Canada's Food Guide.* Ottawa: Minister of Supply and Services, Canada.

Carola, R., Harley, J., Noback, C. (1990). *Human anatomy and physiology.* New York: McGraw-Hill.

CISD Committee, Fire Officers Association of Nova Scotia. (1994). *Critical incident stress in firefighting* [brochure]. Truro, Nova Scotia: Author.

Everstine, D.S., Everstine, L. (1993). *The trauma response treatment for emotional injury.* New York: Norton.

Everly, Jr., G. (1992). Advanced critical incident stress debriefing workshop. Paper presented at the conference of the International Critical Incident Stress Foundation, Halifax, Nova Scotia, December.

Gilliland, B. E., James, R. K. (1996). *Crisis intervention strategies,* 2nd ed. Pacific Grove, CA: Brooks/Cole.

Goleman, D. (1995). *Emotional intelligence.* New York: Bantam.

Hanson, P. G. (1986). *The joy of stress.* Islington, ON: Hanson Stress Management Organization.

Holmes, T. H., Rahe, R. H. (1967). The social readjustment scale. *Journal of Psychosomatic Residence 11*:213-8.

Howatt, W. (1995). *Counselling for paraprofessionals: Formulating your eclectic approach.* Halifax, NS: Nova Scotia Community College Press.

Kuntzleman, C. (1987). *Concepts of feeling good.* Detroit: Fitness Finders.

Lar-Goran, O. (1987). Applied relaxation: Description of a coping technique and a review of controlled studies. *Behavior Research Therapy 25*:397–409.

Lee, M. A. et al. (1985). Anxiety and caffeine consumption in people with anxiety disorders. *Psychiatric Residence 15*:211–307.

Loehr, J., Migdow, J. (1986). *Take a deep breath.* New York: Villard.

Lowen, A. (1969). *The betrayal of the body.* New York: Collier.

Lowen, A. (1975). *Bioenergetics.* New York: Penguin.

Lowen, A. (1989). Bioenergetic analysis. In R.J. Corsini, D. Wedding (eds.), *Current psychotherapies* (4th ed.). Itasca, IL: F.E. Peacock.

Lowen, A., Lowen, L. (1995). IIBA 1994–1995 workshop brochure. New York: International Institute for Bioenergetic Analysis.

Maxwell, W. (1991). Debriefing for staff personnel. CPI National Report. Brookfield, WI: National Crisis Prevention Institute.

McArdle, W. D., Katch, F.I., Katch, V.L. (1981). *Exercise physiology.* Philadelphia: Lea & Febiger.

Mitchell, J. T. (1987, June). *Effective stress control at major incidents* [Maryland Fire and Rescue bulletin]. Baltimore: Maryland Fire Department.

Mitchell, J. T. (1988a). *Defusing.* Halifax: International Critical Incident Stress Foundation.

Mitchell, J. T. (1988b) *Demobilizations.* Halifax: International Critical Incident Stress Foundation.

Mitchell, J. T. (1989). *Introduction to debriefing.* Halifax: International Critical Incident Stress Foundation.

Mitchell, J. T. (1992). Basic critical incident stress debriefing workshop. Paper presented at the conference of the International Critical Incident Stress Foundation, Halifax, Nova Scotia, December.

Murray, M. T. (1995). *Stress, anxiety and insomnia.* Rocklin, CA: Prima.

Orlandi, M., Prue, D. (1988). *Encyclopedia of good health.* New York: Michael Febiger.

Pollack, M., Wilmore, J., Fox, S. (1984). *Exercise in health and disease.* Philadelphia: W. B. Saunders.

Purkey, W., Schmidt, J. (1996). *Invitational counseling:A Self-Concept Approach to Professional Practice.* Pacific Grove, CA: Brooks/Cole.

Rossi, E., Cheeks, D. (1988). *Mind-body healing: New concepts of therapeutic hypnosis.* New York: Irvington.

Rossi, L. (1991). *The 20-minute break.* Los Angeles: Tarcher Putnam.

Selye, H. (1976). *Stress and life.* New York: McGraw-Hill.

Spence, A. (1988). *Encyclopedia of Good Health.* New York: Facts on File.

Walsh, R. (1989). Asian psychotherapies. In R. J. Corsini, D. Wedding (eds.), *Current psychotherapies* (4th ed.). Itasca, IL: F. E. Peacock.

Wells, V. (1990). *The joy of visualization: 75 creative ways to enhance your life.* Vancouver: Rainfrost.

Werbach, M. R. (1987). *Nutritional influences on illness.* Hartford, CT: Keats.

Williams, B. K., Knight, S. M. (1994). *Health for living: Wellness and the art of living.* Pacific Grove, CA: Brooks/Cole.

Wolpe, J. (1969). *The practice of behavior therapy.* Oxford: Pergamon.

Appendix

Student Samples of Counseling Orientations

ECLECTIC COUNSELING MODELS

Here we are at the end of the book. Throughout, we have described a thoughtful "systematic eclecticism," as Egan (1994) calls it: not an arbitrary jumble of techniques but a range of methods that are appropriate to the situation as well as to the general personality and beliefs of the counselor. The opening chapters guided you in choosing an orientation. In this introduction to counseling models developed by students, we simply want to draw your attention to the link between theory and practice.

Counseling psychology tries to define something of the conscious personality that represents each human being. It deals with the "I," the so-called ego, the self that is expressed through one's behavior. Theorists build a counseling method on the basis of their understanding of personality—on some grasp or hypothesis concerning how the "I" functions. The following list, which is by no means intended to do justice to any counseling theory, indicates in a nutshell the relationship between practice and theory.

Jung	The human being (the ego) is connected to a larger reality through the unconscious. Therefore . . . *bring the unconscious into conscious awareness.*
Lowen	The human being is holistic. Therefore . . . *in addition to the unconscious, be aware of how the body feels and functions.*
Rogers	The human being is self-actualizing. Therefore . . . *let the self grow through a warm therapeutic relationship.*
Perls	The human being functions holistically in the present. Therefore . . . *develop awareness of aspects of the present moment.*
Berne	The human being expresses self through its social transactions. Therefore . . . *know how ego states are expressed in these transactions.*
Adler	The human being is socially oriented and chooses its own goals. Therefore . . . *encourage social interest and a new lifestyle.*
Bandura	The human being chooses behavior under the influence of the situation. Therefore . . . *learn to respond differently in certain situations.*
Glasser	The human being chooses behavior according to basic wants and needs. Therefore . . . *help choose behavior that can meet these needs in a responsible and socially beneficial way.*
Beck	The human being chooses behavior primarily by cognitive processes. Therefore . . . *discover whether thinking and assumptions are adaptive enough to work well.*
Ellis	The human being chooses behavior by feelings and especially through its reasoning. Therefore . . . *show whether thinking makes sense.*

Each theory gives an explanation of how the self functions, and each claims to be the best or most significant explanation. These are the *psychological* (and physiological) explanations. Dooyeweerd (1968) points out that there are other (philosophical, historical, or moral) ways to view the ego, and urges us not to absolutize the psychological ego or any of the approaches to it because people are always more than just ego.

Someone may say, "Why choose any? Why not consider all to be right in their own way, or all to be intellectual detours around the practical world of the client?" The answer is that quality counseling is not haphazard. Good counselors know what they are doing and why. They do not treat their clients as objects on the receiving end of a bunch of techniques that keep counselors employed. Though there are many theories to choose from, you must choose a place to start.

The following student orientations, with their accompanying mind maps, incorporate counseling approaches that have been tested by masters in the world of counseling. These students have made a start, and we invite you to review the models as you prepare to formulate your own approach. When you start creating your own orientation, please be sure to take into consideration the PE + PE + CO + O formula on page 8 of this book. Taken together, they are important elements in ensuring your effectiveness as a counselor.

References

Dooyeweerd, H. (1968). *In the twilight of Western thought*. Nutley, NJ: Craig Press.

Egan, G. (1994). *The skilled helper: A problem management approach to helping*. Pacific Grove, CA: Brooks/Cole.

MEREK JAGIELSKI: ADLERIAN, REALITY, AND COGNITIVE-BEHAVIORAL

My chosen counseling orientation is based on three theories: Adlerian, reality and cognitive-behavioral. In my opinion, all of them share the cognitive approach to the individual's behavior. They are action-oriented, structured, and present-centered, and allow the counselor to employ numerous cognitive, emotive, and behavioral techniques. They also emphasize effectiveness, quick results, and a counselor's inventiveness in providing the client with learning tools. Their focus on teaching and learning, with no limitations on a counselor, have convinced me that my teaching experience could be used effectively. As a result, my counseling orientation has become an integrated and eclective model.

My relationships with the client are based on full acceptance, friendly tolerance, and collaboration. I believe that such a defined relationship will protect me from getting strangled by a client's need for emotional dependence during therapy. I view my counseling role as assisting a client who want to change from getting stuck in behavioral or emotional status quo. I place emphasis on the client's autonomy and his willingness to cooperate during therapy. I respect a client's capacity to influence, interpret, and create events, and regard this ability as crucial in behavioral change.

My therapy goals are aimed at maladaptive behavior. This behavior is a product of an individual's motivation distorted by mistaken goals, faulty assumptions, irrational thoughts, automatic thinking schemata, and present needs or wants. This results in disorganized and deprived-of-purpose behavior. The primary focus of the therapy goals is put on present *here and now* behavior, including internal determinants such as values, beliefs, attitudes, interests, and individual perception of reality.

My therapeutic techniques are basically oriented toward exploring, educating, reorienting private logic, and addressing irrational thinking that affects motivation. By providing information, teaching new skills, guiding, and challenging, while using cognitive, emotive, and behavioral techniques, the counselor introduces clients to new understanding of their behavioral dimensions.

The client's life experiences shape and affect present thinking and motivation. Life assessment is necessary for better insight into origins of maladaptive behavior. This assessment deals mostly with a client's experiences so as to find a cause of irrational thoughts and faulty assumptions. It also answers questions: *Where do these attitudes come from? How do they relate to present behaviors and the client's perception of self?* This influences clients' awareness of their origins and helps them to identify their negative impact on behavior. The change of a client's attitude toward the past may become a driving force for self-evaluation and an encouragement to work collaboratively on behavioral alternatives during therapy. In this reorientation phase, the client is encouraged to make a distinction between subjective and objective frames of reference. This is done by using: acting as if, push a button, empty chair, reframing, and genogram techniques.

The educational aspect of this part of therapy focuses on pointing out negative consequences that earlier experience might have on present behavior.

The client's irrational beliefs greatly affect his behavior—especially thinking and acting. This results in developing maladaptive behaviors perpetuated by repeti-

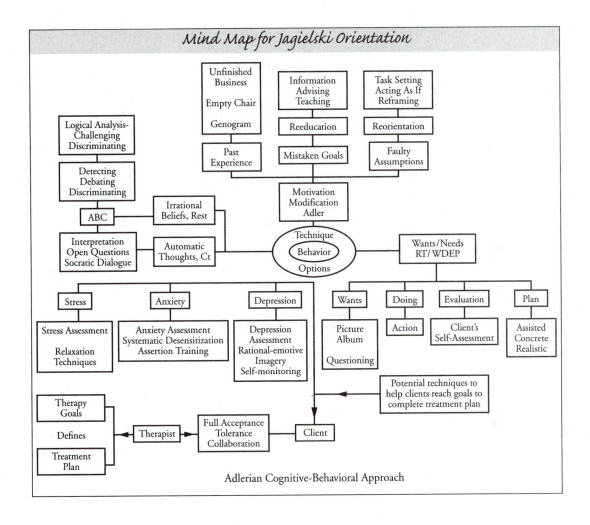

Mind Map for Jagielski Orientation

Adlerian Cognitive-Behavioral Approach

tive actions based on the same irrational premises. The point is to show the client that they have no relevance to objective reality. I have included the ABC model in my orientation to deal specifically with these problems. It serves to teach clients to be responsible for creating their own emotions and following their behavioral responses.

The client's present needs and wants are also given consideration. This focus explores present behavior, with a client's immediate wants and needs included, to help the client to become more self-motivated through seeing that change is possible and attainable. Clearly defining with clients their wants and transforming them into an effective plan of action, based on WDEP (Wants Doing Evolution Plan) technique, can stimulate them to take care of personal behavior and act responsibly.

The WDEP technique helps clients recognize and define what appropriate behavior they want to choose to satisfy their needs. Through WDEP stages, clients learn that it is up to them to become more directive in their own behavioral change. This action-oriented technique forces a client to do something, and prevents co-dependency. It also makes participation in the therapy more active and dynamic.

I assume that there always will be some kind of "behavioral emergency," which I identify as stress, anxiety, and depression. I intend to use emotive and behavioral cognitive techniques as working tools to deal with stress-related problems.

My counseling orientation is integrative and eclectic, directed mostly toward a change of behavior. It may be applied to adolescents and to cultural conflict counseling, which is my area of interest. All of the approaches are interwoven, and can be used interchangeably, depending on the therapeutic situation and goals. The designed form of my orientation leaves room for additional modifications as my understanding grows. For the time being, it illustrates my basic approach toward a client's maladaptive behaviors and gives me the flexibility to use any techniques appropriate to a client's priorities, my own assessment, and the therapy goals.

CASE STROMENBERG: COGNITIVE, ROGERIAN AND CHRISTIAN

1. Begin
   A. Set the environment
      - Be friendly and open
      - Speak as an equal who is qualified to help; be deliberate but not hasty (Collins)
      - Show empathy and acceptance (Rogers)
   B. Collect background data
      - Engage in casual conversation about personal interests, preferences, vocation/hobbies, home and family, social interests (clubs, sports, church)
      - Show interest
      - Listen both for factual information and how it is said
   C. Minimize surprises
      - Are you seeing (or have recently seen) another counselor?
      - Discuss policy of confidentiality and breaking confidentiality
      - Discuss fee payment, if any; number of sessions subsidized, if any
      - Explore expectations: What would you like from our session?
      - Give a very brief explanation of your orientation or method, if requested
2. Define
   A. Hear concerns
      - What brings you here today? What is on your mind?
      - Use active listening skills; reflect and summarize (Rogers)
   B. Explore functional levels
      - Briefly explore modalities of concerns (Dooyeweerd, Olthuis, Mowrer)
         (1) the physical (medical-biological; medications)
         (2) the psychological (behavior)
         (3) the productive (work, achievement, creativity)
         (4) the relational (personal and social relationships)v. the responsible (conscience, character)
         (5) the religious (meaning, purpose, acceptance)
      - Get a sense of the degree of hope/anxiety in regard to self, one's situation, the future (cognitive triad; depression inventory questions) (Beck)
   C. Define and/or prioritize concerns
      - Summarize for client, then ask which concern levels need to be dealt with right away, or which is the most pressing
      - Focus the concern: What is it you really want in this? (Glasser) What bothers you most about the situation?
      - If there is more than one major concern, decide together the order in which they will be addressed
3. Consider
   A. Reduce immediate anxiety
      - Behavioral techniques: deep breathing, muscle stretch and relaxation, walk or change position, background music, guided imagery (Wolpe, etc.)

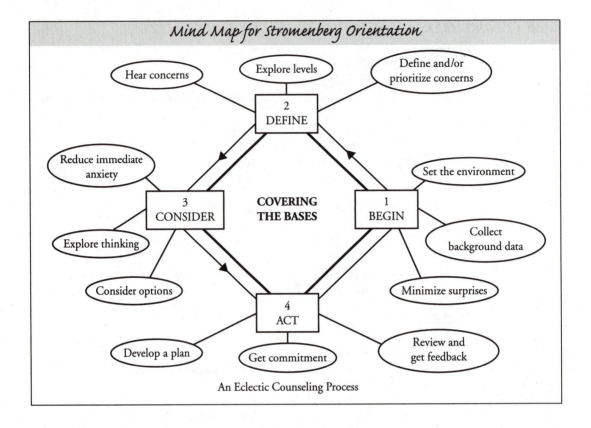

Mind Map for Stromenberg Orientation

An Eclectic Counseling Process

- Cognitive techniques: generate alternatives/options, decatastrophize, re-attribute, redefine, decenter (Beck, Bandura, Meichenbaum)
- Other: magic wand (Adler); emphasize present security, crises are temporary, available help; stability of basic hope and beliefs (Adams); itemize available options regarding material concerns

B. Explore thinking
- Relate behaviors and psychological needs to personal choice, affect (feelings), and cognition (Glasser, Bandura)
- Collaborative discovery, Socratic dialogue: What would you be thinking when. . . ? Let's examine this belief . . . ; Suppose that . . . ; How does that work out? (Beck)
- If necessary, teach human functioning: relationship of thinking/feeling/behavior (Beck); relationship of behavior to conscience and beliefs (Mowrer); relationship of feelings to chosen behaviors (Adams)

C. Consider thinking/activity options
- Have clients evaluate own behaviors and present thinking; discuss the basis of this evaluation (Glasser)

- Use this self-evaluation as a springboard to consider other thinking and options
- Explore options: read briefly from appropriate material, use visual aids

4. Act

   A. Develop a plan
   - How can you get to where you want to be?
   - Explain progression of vision-plan-effort-achievement
   - Choose an optimal plan; consider a few of its challenges (i.e. how its difficulty may be overcome)
   - Divide the challenge into practical steps

   B. Get a commitment to change
   - Are you ready to. . . ? What do you need to be ready?
   - How long will you give yourself to solve this? What will you do about this today/tomorrow/ in the next three days?
   - Give an appropriate homework assignment (Bandura, Beck, Adams)

   C. Review and get feedback
   - Review highlights of the session
   - Allow client to express feelings and thoughts about it: How has this session gone for you? (Beck)
   - Discuss further contact

References

Adams, J. E. (1973). *The Christian counselor's manual.* Phillipsburg, NJ: Presbyterian and Reformed.

Adler, A. (1963). *The practice and theory of individual psychology.* Paterson, NJ: Littlefield, Adams.

Beck, A. T., Rush, A. J., Shaw, B. F., Emery, G. (1979). *Cognitive therapy of depression.* New York: Guilford.

Collins, G. (1988). *Christian counseling: A comprehensive guide,* revised. Dallas: Word.

Dooyeweerd, H. (1957). *A new critique of theoretical thought.* Grand Rapids: Wm. B. Eerdmans.

Meichenbaum, D. (1977). *Cognitive-behavior modification: An integrative approach.* New York: Plenum.

Mowrer, O. H. (1966). *Abnormal reactions or actions.* Dubuque: W.C. Brown.

Olthuis, J. H. (1994). God-with-us: Towards a relational psychotherapeutic model. *Journal of Psychology and Christianity 13:* 37–49.

Rogers, C. R. (1951). *Client-centered therapy.* Boston: Houghton Mifflin.

Wolpe, J. (1958). *Psychotherapy by reciprocal inhibition.* Stanford, CA: Stanford University Press.

Wubbolding, R. E. (1988). *Using reality therapy.* New York: Harper & Row.

MARK BISHOP: ROGERIAN AND COGNITIVE

1. Contact
   A. Set therapeutic environment
      • Rapport-building microskills
      • Matching/mirroring
      • Breathing
      • Anchoring
      • Voice tonality, rate, volume
   B. Rogers' core conditions
      • Congruence (be honest and open)
      • Empathy (sensing and perceiving client's subjective world)
      • Unconditional positive regard (leading, through modeling, toward clients' unconditional self-acceptance)
   C. Define therapeutic relationships
      • Address and discuss with client the following areas:
         (1) Confidentiality
         (2) Probable number and cost of sessions
         (3) Your counseling orientation, techniques, therapy procedure
         (4) Client's background information
         (5) Expectations, fears, assumptions, feelings about therapy
   D. Assess clients' willingness and readiness to change according to where they are on the continuum from precontemplation, to contemplation, preparation, action, maintaining, or termination
      • Examples:
         (1) What do you want from counseling?
         (2) Are you willing to do what is necessary to change?
         (3) Do you want to change?
   E. Develop treatment plan based upon goals set by the client; however, if you are unable to help client, refer and support
2. Explore
   A. First reduce anxiety associated with change and perception of self; particularly in terms of overcoming a problem and achieving a goal.
      • Relaxation exercises (ha breathing, muscle relaxation, guided imagery)
      • Decatastrophize (use of what-if situations): What if you didn't visit your mother, would she die?
      • Decentering (dissuade clients that they are the world's focus): How many people are actually paying attention to you?
      • Reframing (put situation in a different light so client can perceive it differently): When you say everyone would hate you, don't you just mean you? (adapted from Beck)
   B. Questioning and client self-exploration to examine core beliefs/thoughts through:
      • Role play, where a hypothetical, typically problematic situation is enacted in order to gauge a client's thoughts and actions/reactions to the perceived situation (Beck)

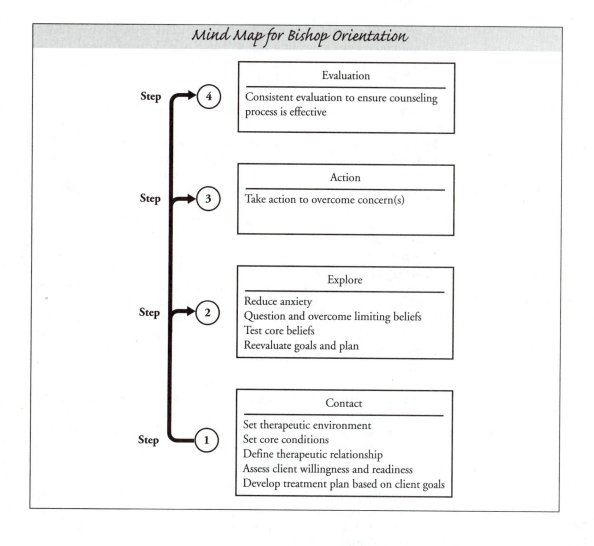

Mind Map for Bishop Orientation

Step → ④

Evaluation

Consistent evaluation to ensure counseling process is effective

Step → ③

Action

Take action to overcome concern(s)

Step → ②

Explore

Reduce anxiety
Question and overcome limiting beliefs
Test core beliefs
Reevaluate goals and plan

Step → ①

Contact

Set therapeutic environment
Set core conditions
Define therapeutic relationship
Assess client willingness and readiness
Develop treatment plan based on client goals

- Scale technique, where therapist charts and explains thought/feeling/action connection (Beck)
- Explore meanings, in order to determine what meaning a situation has to client, and thoughts and reverberations experienced with it (Beck and Weishaar)

  Through this exploration, the counselor formulates ideas of a client's cognitive pattern and what needs to be ameliorated while clients learn the connection between thoughts and feelings, and that they are in control of, and at cause for, cognitions and their subsequent feelings.

C. Test perceived core beliefs to determine accuracy and lay groundwork for change. This can be accomplished via:

- Socratic dialogue, or open-ended questioning with clients, where they ultimately form their own conclusions (Beck)

- Behavioral experimentation (Beck). This technique tests the client's faulty assumptions by testing them against empirical evidence of reality.
  (1) "My mother yelled at me today, she hates me." "Does she always yell at you" and/or "Do people ever yell at people they love?"

D. After client and therapist have had an opportunity to examine and explore the cognitive process and subjective world of the client, re-evaluate goals and plan.

3. Action

A. New, adaptive, and healthy ways of thinking are attempted after the client has approved the newly set and evaluated goals and made a commitment to achieving them. This can be attempted through exercises such as:

- Behavior rehearsal, where client performs tasks in a safe setting before trying them in the real world
  (1) Say hi to the therapist before saying hi to a stranger
- Shame-attacking, where individuals place themselves in a situation they perceive as hostile and where they think they will fail, in order to test this faulty hypothesis and prove it wrong.
  (2) Approaching and talking to 100 beautiful women while thinking they will always be rejected (Weidinger and Crookson-Rose, 1997)
- Bibliotherapy and homework exercises are used to further facilitate the learning and adapting process, and empower clients with thought that they are in control of, and responsible for, their own therapy and ultimate well-being.

4. Evaluation

A. In order to determine progress for client *and* therapist, each exercise and session should be evaluated, both by asking the client, and by using criteria of homework, completed tasks and exercises in-session, and counselor's own interpretation. In addition to giving the client and therapist a sense of progress, this also enables the therapy to adapt, either to new goals or priorities or to crisis situations.

B. When either the goals and objectives are met or the sessions are completed, the therapeutic relationship is terminated, with the therapist suggesting further bibliotherapy and other resources for the client, and scheduling a follow-up for the future.

ANGELIA GRAVES: REALITY, ADLERIAN AND BEHAVIORAL

1. Reality therapy (Glasser, Wubbolding)
   A. Set the environment
      - Develop friendship and trust
      - State confidentiality limits (e.g., what you say to me is kept confidential, except if you tell me that you plan to harm yourself or others. I am then legally obligated to break confidentiality)
      - Ask about other professionals (Are you seeing another counselor? A lawyer or doctor?)
      - Ask what the client expects from counseling
      - Explore needs assessment with client (family, friends, job, school)
   B. Client's wants, needs, and perceptions. Use open-ended questions to explore all aspects of clients wants, needs, and perceptions and those of family, friends, etc. (What do you want? What do others want from you? What do you really want? How do you look at it? What does that look like? If you had what you wanted, what would that give you?)
   C. Present behavior of client
      - Explore clients' present behavior and inquire whether it is helping them get what they want (What are you doing? Is it helping you? What would you like to be doing? How has it been working for you? Tell me about your day: what did you do?)
   D. Total behavior of client
      - Explore client's thinking, feeling-and-doing behavior (What are you doing? What were you thinking when . . . ? What were you feeling when . . . ?)
   E. Develop a plan with client
      - The plan is based on the client's needs and wants. It should be realistic and attainable (Are your wants realistic? Can you attain that?). Examine the pros and cons of the plan (What if this doesn't happen? What are some other options if this doesn't work out?). Gain commitment from client (When will you begin this? Where?)
2. Adlerian (Corsini and Wedding)
   A. Act *as if* to practice change (Describe to me someone with confidence. Act as if you have confidence.)
   B. Social contact
      - Develop it with the client (Are you involved in any groups? How does your behavior affect others?)
3. Behavioral (Todd and Bohart)
   A. Explore the personal and situational factors relating to the client's problem (Where? When? What?)
      - Examine client's self-efficacy and help to develop self-confidence
      - Possibly have clients make list of their strengths, talents, or positive attributes, then pick one or more to focus on. Client reminds self each day (e.g., through carrying a positive message on a small card)
   B. Teach assertion and communication skills through role playing and rehearsal

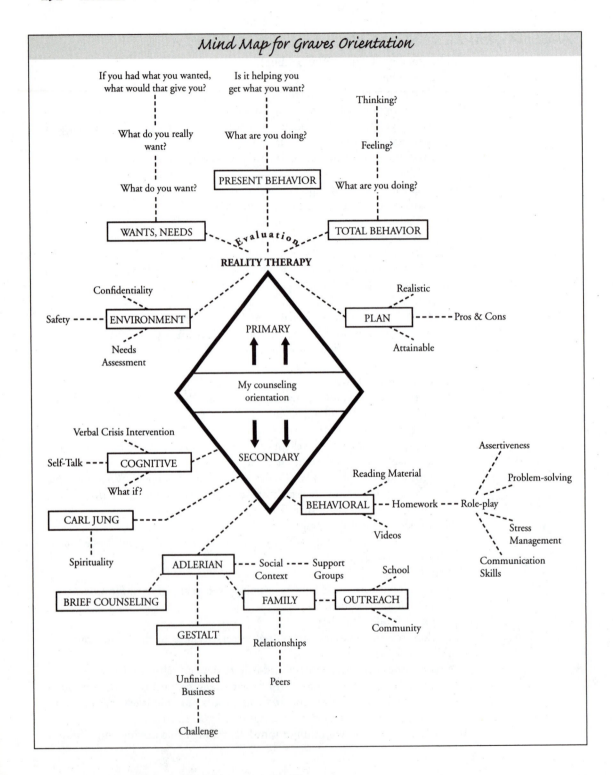

Mind Map for Graves Orientation

If you had what you wanted, what would that give you?

Is it helping you get what you want?

Thinking?

What do you really want?

What are you doing?

Feeling?

What do you want?

PRESENT BEHAVIOR

What are you doing?

WANTS, NEEDS

Evaluation

TOTAL BEHAVIOR

REALITY THERAPY

Confidentiality

Realistic

Safety

ENVIRONMENT

PLAN

Pros & Cons

Needs Assessment

PRIMARY

Attainable

My counseling orientation

Verbal Crisis Intervention

SECONDARY

Assertiveness

Self-Talk

COGNITIVE

Problem-solving

What if?

Reading Material

BEHAVIORAL — Homework — Role-play

CARL JUNG

Videos

Stress Management

Spirituality

ADLERIAN — Social Context — Support Groups

School

Communication Skills

BRIEF COUNSELING

FAMILY — **OUTREACH**

Community

GESTALT

Relationships

Unfinished Business

Peers

Challenge

C. Guided imagery
- Client imagines situation (e.g., if client fears speaking in front of an audience, say: "Imagine yourself in front of an audience. What are you doing, thinking, feeling?")

4. Cognitive (Todd and Bohart)

A. Self-talk (Meichenbaum)
- Practice positive sayings to yourself each day (I can do this!)

B. Graduated task assignment (Beck, 1960)
- Client begins an activity at a comfortable level; the difficulty of the task is gradually increased (e.g., if clients are shy, let them begin by saying hello, and gradually move toward starting a conversation)

C. Socratic dialogue with client
- Let us consider the steps of what you did. Where did things go wrong? What did you find helpful or not helpful?

5. Family Counseling (Todd and Bohart)

A. Open systems (Bowen)
- Look at the way the client's behavior affects the family or others (How does your behavior affect your spouse/ children/ friends? When you . . . , how do others react?)

B. Role reversal
- If a couple has distinct roles, have each play the part of the other

6. Peer counseling (Muro and Kottman)

A. Support the client with someone who has commonalities (I would like to put you in touch with someone who has experienced the same thing you have. Would you be interested?)

References

Corsini, R. J., Wedding, D. (Eds.). (1989). *Current psychotherapies* (4th ed.). Itasca, IL: F. E. Peacock.

Muro, J., Kottman, T. (1995). *Guidance and counseling in the elementary and middle schools.* Dubuque: Brown and Benchman.

Todd, J., Bohart A. C. (1994). *Foundations of clinical and counseling psychology.* New York: Harper & Row.

Wubbolding, R. (1988). *Using reality therapy.* New York: Harper & Row.

JANET SEARS: ROGERIAN, ADLERIAN, JUNGIAN AND COGNITIVE

1. Rogers' core conditions (Corsini and Wedding, 1988). Establish a safe environment for the client through practice of Rogers' core conditions:
   - Active listening—paraphrasing and reflecting back to client
   - Practising unconditional positive regard—valuing the client regardless of what is disclosed (This does not mean unconditional acceptance!)
   - Practising empathy and acceptance
   - Being yourself—being genuine within the professional boundaries of your counseling role

2. Adler's family constellation (Corsini and Wedding, 1998; Todd and Bohart, 1994)
   - Exploration of family constellation: data gathering on family of client, client's relationships with parents, siblings, spouse, children
   - Exploration of client's social interests: data gathering on friendships, social activities, community interest
   - What client is giving and/or receiving from society: data gathering on career, volunteer activities, interpersonal relationships
   - Exploring whether client is experiencing discouragement in pursuit of social interests

3. The mind
   A. Reality therapy (Wubbolding, 1988)
   - Needs assessment: what the client wants, what others want from the client; explored in the context of family, friends, employers, acquaintances
   - Reality check: explore whether what client wants, what others want from client, is realistic and attainable. Also explore whether these wants are acceptable to the client, to others, and to society

   B. Cognitive techniques (Gladding, 1991)
   - Decentering, redefining, magic wand (Adler), role playing, brainstorming (Egan), assigning tasks to encourage change in thinking; for example, journal writing (tracking feelings), use of bibliotherapy, use of self-help and support groups, providing new information on relevant topics such as self-talk, etc.

   C. Behavioral techniques (Todd and Bohart, 1994)
   - Relaxation techniques: breathing, use of music, physical exercise (walking, sports), guided imagery

4. The planning process (Egan, 1994; Wubbolding, 1988)
   A. Together with the client, create a plan: it should focus on the present life experience of the client and involve "doing." Factors to be discussed include:
   - What
   - Where
   - Frequency
   - Client evaluation of the plan (whether it is helpful)
   - Attainability and acceptability of the plan
   - Client's self-efficacy (degree of confidence) in relation to the "doing" component of the plan
   - Seek agreement by the client to commit to the plan (Egan)

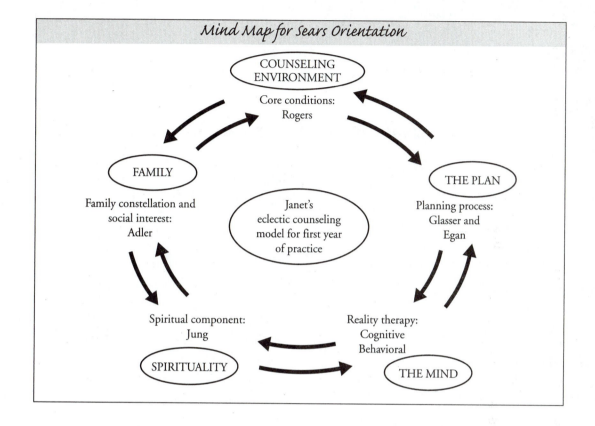

Mind Map for Sears Orientation

COUNSELING ENVIRONMENT

Core conditions:
Rogers

FAMILY

Family constellation and
social interest:
Adler

THE PLAN

Planning process:
Glasser and
Egan

Janet's
eclectic counseling
model for first year
of practice

Spiritual component:
Jung

Reality therapy:
Cognitive
Behavioral

SPIRITUALITY

THE MIND

B. Follow up in future counseling sessions by reviewing the plan, asking for feedback, and having the client evaluate how the plan is working

5. The spiritual component (Corsini and Wedding, 1998). This may be discussed with client as requested, as an option for the client, and/or as a vehicle of hope and courage

   • Explore with client the degree of meaningfulness of present spiritual experience
   • Various spiritual options in the community may also be discussed
   • Use of bibliotherapy may generate greater awareness, knowledge, and self-growth

References

Corsini, R. J., Wedding, D. (Eds.) (1989). *Current psychotherapies* (4th ed.). Itasca, IL: F. E. Peacock.

Egan, G. (1994). *The skilled helper: A problem management approach to helping* (5th ed.). Pacific Grove, CA: Brooks/Cole.

Gladding, S. (1991). *Group work: A counseling specialty.* New York: Macmillan.

Todd, J., Bohart A. C. (1994). *Foundations of clinical and counseling psychology.* New York: Harper & Row.

Wubbolding, R. (1988). *Using reality therapy.* New York: Harper & Row.

INDEX TO TOOLBOX TECHNIQUES

General Index

William A. Howatt

William (Bill) Howatt has been involved in the field of counseling and human performance for the past fourteen years, as a teacher, counselor, and coach. He has offered services in counseling psychology, and EAP, and is now focused on human resources and personal performance life coaching.

Bill Howatt is primarily known for his counseling expertise in the area of addictions, trauma, anxiety, and stress. He is now exploring the impact of emotional ergonomics in the workplace. Most recently, Howatt has been involved in the personal performance of professional athletes. He uses his own books, such as *My Personal Success Coach, Golf Psyche, An Employee's Survival Guide for the Twenty-First Century, A Parent's Survival Guide for the Twenty-First Century, My Personal Change Journal,* and *Journaling My Journey,* to facilitate his personal success coaching.

In the area of human resources, Bill Howatt has developed communications training, employee screening, climate and team building, and an executive coaching program. He teaches programs in the areas of education and counseling for several universities and **Nova Scotia Community College.** To assist with the delivery of his curriculum, he has been very active over the past few years, publishing eighteen books and several articles for journals, and he is a frequent speaker at conferences on staff development in the area of human behavior.

Bill Howatt has extensive training in the counseling field. He holds a Ph.D. in counseling psychology; is an advanced practicum instructor with the William Glasser Institute; has advanced training in thought field therapy; had advanced training in critical incident stress; is a master practitioner in neurolinguistic programming; is a master practitioner in time line therapy; is a master hypnotherapist; has advanced training, and is certified in, handwriting analysis; holds advanced certification in alcohol, drugs, and gambling addictions; and has advanced training in conflct resolution.

Credits